How to Live Through a Famine

Dean L. Rasmussen

Third Edition,
Revised and Enlarged

To my loved ones

CONTENTS

Preface 1

Orientation 3

Survival as an individual 7

Edible plants

Color plates

Insects for food

Survival as a group 125

Use of land

Fibrous proteins as a food source

Plant fiber as a food source

Chemical fractionation of plants

Microbiological cultures

Bibliography 139

Index 143

PREFACE

The purpose of this book is to furnish information that will help a population to survive the advent of famine. It would be a waste of time to write a book on survival, when so many have already been written. As opposed to helping perhaps one or two people out in a wilderness, this book is specifically concerned with survival on a population-wide basis.

The need for this approach should be clear. After shelves become bare in a famine, people will tend to struggle in a frenzy for whatever they can find to eat. An individual might have a trusty manual that says cattails and watercress are edible. So he faithfully travels out to some cattails—former cattails, to find that 700 people had got there first. He might hike over to gather a patch of watercress, which another person would say he saw first. The winner might be the better gunman. The value of the book lies in its presentation of alternative food sources, where a human being has no other place to turn. Outside of some native plants, there is probably little in these pages that is more delicious or desirable than what people are accustomed to eating. They are just alternatives. No ordinary book on survival would indeed need to mention these options.

When there is famine, what will happen? Are some of us destined to repeat history and die of starvation?

When we consider the options open to us, there seems to be no way that this can happen. There are so many alternate sources of food, and these sources are so potentially abundant, that no one really has to starve to death. Yet man seems to have a profound capacity for botching things up, and individuals and groups will probably die. Nevertheless, there is no reason for this to happen to us, if we are prepared for the event when it comes. To this end this book is written.

ORIENTATION

One should realize at the outset that there is no way to magically transform a person or group into readiness for a famine who have no training in areas related to it. If one's abilities are confined to how to open cans or packages of frozen foods, then he is simply not prepared for what is to come. Individuals or groups, to live through a famine, will have to feel sufficiently motivated to learn what they do not know related to the subject. This will mean literacy, in particular, in chemistry and in life sciences, such as identification of plants and microbiology.

Bound by limits of time and money, the author has emphasized a broad approach to the subject, where everything is visible in proper perspective. The alternative approach would have been to spend years compiling detailed information on just one or two topics, which is an approach open really only to a group with resources in time and money. All that can be done here is to provide insight into the problem and to provide enough details for the reader to intelligently proceed.

We should realize that the measures we are to take during a famine are in part dependent on what will have caused it. If plant life in a region is seared off the face of the earth from a high altitude nuclear blast, it will not do much good to talk about foraging for edible plants. If starving mobs surge through an area every ten days, it might be quite pointless to talk of raising normal crops in that area.

Let us commence, therefore, by discussing the train of events which will occur in any kind of famine, whether caused by anarchy, drought, or whatever. Things will start off with a latent period of undefined length during which the shock of the fact will set in. No markets, no restaurants, no fruit stands, no drugstore candy racks. People will search their alternatives. There will be a sizing up of what one has and how far one can go.

As food stocks run out, the hard facts of life will begin. There will be varying degrees of violence toward whoever is thought to still have food. There will be wholesale breaking and entering, and an increasingly animal search for anything to eat. Such conditions would be more the case in America, where people are well armed, than somewhere like former Biafra. With Americans thus armed, no police force could hope to contain them– even if they were well fed and on the job.

At all levels, there are common sense procedures to follow, and they in fact must be followed, if anyone is to escape the hell of famine. For an individual who has no alternatives, he should scatter what food he has, wherever he can. If one has a sack of flour and six cans of carrots, he should put each can in a different place. He should deposit the flour in whatever containers he can find– pop bottles, shoe boxes, coat pockets, anything suitable. If such a dwelling is raided, the owner may indeed lose some of his cans and flour. But they probably will not get all six cans, or the whole bag of flour. If the same person is cut off from all food, he can still count on living for quite a while. The report was made of one man who was so fed up with things that he decided to take his life. The method he chose was to no longer eat. At the end of 72 days without food, he was still around, and he was getting quite nervous because of the fact that he was still hanging on. Finally he figured to himself that he was not supposed to go yet, so he returned to normalcy.

One's chance of survival through a famine will be most directly related to whether or not he is a member of some group, and the size of that group. By group, we must understand a moral unit, one where its members act responsibly toward one another, where they will sacrifice if necessary for each other, where they will not steal food from each other, and so on. If this group is twice as large as another group, it has probably somewhere near twice the chance of helping its members to live. There is a critical advantage in this thing of numbers– one's life may well depend on the group being large enough.

The first reason for being in a group is for defense. One person cannot guard belongings on a 24 hour basis; two or more people can. Still more guards provide better surveillance and lessen the chance of being overwhelmed when marauders come.

If the group is large enough, this means there can be enough people to run one or more critical processes on a continuous basis. Others can gather raw materials; others can protect them. There will be enough raw power in the group to collect and protect equipment.

Most important of all, a group will have the ability to police an area and protect it, corresponding directly to that group's size. It can control who leaves and enters, and what they do inside.

The point should in fact be made that without groups exercising such control over natural areas, it could mean the end of some native edible plants. Pine trees could cease to exist, through the graces of mobs cutting them down to get at the inner bark. Cattails are too well known to escape the same fate.

Who will these groups be? Where will personal loyalties settle? Things will center around stable, moral groups, such as church and family units. Organized church groups would be better than family units, for the simple advantage of sheer numbers. In the same way, an entire church organized would be more effective than preparedness within random groups of that church. It is open to question whether federal or state governments would be valid groups in such a situation. In modern American government, there seems to be an appalling lack of moral conviction or ethics. And these would seem to be precisely the qualities which would determine the durability of a group during famine. To say that famine would cause a fall in government might be too much of an oversimplification. This would depend on many factors, among them the governments' ability to commandeer and the completeness of the famine.

If there is still enough time, what should be done? What should you as an individual do? The need for preparation is urgent. Membership in groups should be verified, food stocks should be built up, and means of defense affirmed. Group boundaries should be defined and contingency plans made. The food processes described here should be prepared for, and alternatives explored. More research should be conducted. People should wake up. They should realize that guarantees have been made that famine will come. If after all the warnings have been made, people continue to dawdle, one can only conclude that they were tired of living.

SURVIVAL AS AN INDIVIDUAL

For the purposes of this work we shall assume the presence of plant life. (If we were indeed to suffer from a high altitude nuclear attack, the absence of natural plant life would deny us around 60% of the options in this book.)

The contents of this section are devoted to measures that an individual can apply toward feeding himself in famine. One of his best hopes for survival will be the hunt for native edible plants. If he is successful enough, there is no class of nutrient which he shall not enjoy. Compared to some other areas, identification of plants is quite easy. Plant recognition is fairly simple to learn, and costly equipment is not required for identification (as with micro-organisms). The comprehensive nature of the listing to follow has precluded for the present any chance of describing each entry. Other than the plants which are described in the illustrated part of this book, a wider treatment of what these plants look like is available in the excellent works mentioned opposite page 54.

What follows, then, is a listing of native plants which have been reported as being edible under certain conditions.

Except for common food plants like carrots or peas, this list has been made as comprehensive as reasonably possible. Surely one faced with famine will appreciate knowing about the greatest number of potential food sources. There are undoubtedly many edible plants not in the list, so it would be wrong to regard it as the last word. Even so, it may well be the most comprehensive list of wild edible plants that exists as of this writing. The plant parts in each entry and their methods of preparation are not necessarily the only acceptable ones, or they may be. If there is no poisonous note under a plant, that does not mean it is not poisonous; but notes on the toxic nature of the plants have been included where possible.

The comprehensive nature of the listing may be further

justified where the reader has few or none of the listed plants to draw upon. Where one has no choice, he may cautiously try relatives of the plants listed. Thus one may have *Amaranthus fimbriatus* growing in his yard, which has not been reported as being edible. Since other species of *Amaranthus* are in this list, *A. fimbriatus* may indeed be edible. Or perhaps not. Here at least is another possibility. The old plant on the windowsill may be more than just attractive.

It may come to the point where it is worth trying a plant one knows nothing about. These rules should then be applied:

1. Without swallowing, chew a small part of the raw plant. Remove it, then wait for something that signals danger, such as burning, stinging, or numbness of the mouth.
2. If this happens, cook it; heat destroys many plant toxins. Repeat the above step. If there is still an unpleasant sensation, the plant part should be avoided.
3. If there is no danger signal from a plant that is raw or cooked, then consume a small amount. Wait an hour or more. (The longer the wait the better. Some symptoms of poisoning are well delayed, such as with *Amanita phalloides*—nothing may happen for as long as 15 or more hours. With liver poisoning, symptoms may not surface for days.)
4. Gently increase the amounts eaten. Space them well out until you are sure it is safe. It would make sense to work it in with familiar food, if there is any.

Even with the plants listed, it is well to bear in mind that they are most likely new, and one should first be cautious in their use. As mentioned previously, poisonous qualities of these plants have been pointed out where known. It is true that people such as Indians did not necessarily base their eating habits on this type of information, but some of them were not noted for living long lives, either. It is best to proceed with care and respect where warnings are listed. (Many plant poisons do not accumulate in the body, and where this is fortunate enough to be the case, moderate use of the plant is harmless. A good example of this is nitrates. Although they are toxic, this does not prevent man from enjoyment of celery, radishes, carrots and lettuce, all of which have been shown to contain nitrates. It should be clear that very much hinges on whether a poison is cumulative or noncumulative.)

Eating raw plants also presents certain hazards worth pointing out. Some insects spend part of their life cycles on things like leaves, and may be detrimental if eaten. An example of this would be the liver fluke (*Fasciola hepatica*), which may spend part of its life cycle on the leaves of aquatic plants, such as watercress. Eating such plants raw could make the careless consumer a host to these liver flukes. Another hazard is presented by man's penchant for spraying for everything he decides he does not like. Leaves with insecticides on them might be of questionable value to a hungry person. The best approach with using raw plants for food is to know where possible of specific hazards, and to meticulously wash the plants and inspect them before use.

Where there is a choice of foods, it will pay in the long run to remember that the use of a root or perhaps the inner bark of a plant means its destruction, with a cutoff in its food making capacity. Thus it is preferable to use the seeds or something like part of the leaves of a plant, where it is not destroyed. If it is an annual or biennial that has had a chance to reproduce, then this would not matter.

To avoid confusion, the plants are listed by their scientific names. The numbers below each entry refer to sources given in the bibliography.

Abies spp. Fir species (2)	The inner bark can be eaten raw or boiled. The sap is worth gathering in late winter or early spring when it is flowing.
Abronia latifolia Seacoast abronia (24)	Chinook Indians have eaten the long, stout roots.
Abrus precatorius	The seeds are boiled long enough to detoxify them.

Precatory bean,
rosary pea

One of the most poisonous substances known, abrin, is present in the beans or seeds. Kingsbury reports in fact that less than one seed is sufficient to kill an adult human being. Abrin, however, is protein in nature, and is destroyed by sufficient heat. Under normal conditions (such as moderate elevation), boiling them for ten or so minutes should be sufficient to accomplish this.
(4, 19)

Acer glabrum
Rocky Mountain maple
(16)

The shoots and inner bark have often been of value in emergency situations. The shoots of this plant can be prepared like asparagus.

Acer rubrum
Red maple

Eating the leaves of red maple has caused livestock fatalities, which should be borne in mind if one is inclined to try other parts of the tree fòr food.
(19, 24)

Acer spp.
Maple species
(15, 16, 24, 25)

The winged seeds can be roasted for food. Sap from any *Acer* can either be used as a refreshing drink or it can be boiled down into syrup. Harrington reports the bark of *Acers* as edible, without specifying the inner bark.

Acorus calamus
Sweet flag,
calamus
(8, 15, 24)

The unborn leaves are excellent for salad in the springtime. The rhizome can be dried to remove some of its pungent quality, then chewed or used for tea. The rhizomes can also be prepared into candy (by parboiling to reduce the pungency, then boiling in syrup), but the underground part of the stem is better for this purpose.

Aesculus californica California buckeye

Seeds should be prepared for consumption in this way: Take ripe ones only, roast, crush and then leach them. This process of leaching must be thorough. Sweet says this can be best achieved by letting water run over the meal for ten days. The seeds can also be prepared by burying them in swampy, cold ground during the winter, to remove the bitterness from them, and then boiling them in the spring.

The flowers of this plant and the honey produced from them are reported to be poisonous to humans, as well as other parts of the plant. The nuts of other members of this genus are definitely poisonous and have caused deaths. Caution would be recommended in the use of this plant. (19, 33)

Agaricus arvensis Field mushroom (15, 24, 35)

A variety of this is the common mushroom sold in markets. The wild variety can be prepared in any of the same ways.

Agaricus augustus (32)

With mushrooms that are unfamiliar, it is a standard procedure to always cook them well.

Agaricus campestris (15, 24, 32, 35)

This can be prepared in any way, as with *A. arvensis.*

Agaricus crocodilinus

Unfamiliar mushrooms should always be well cooked.

A. pattersonae
A. rodmani
A. silvaticus
A. subrutilescens
A. sylvicola
(32, 35)

Agastache foeniculum
Fragrant giant hyssop
(24)

Missouri valley Indians used this as a flavoring in cooking, and for making a drink.

Agastache urticifolia
Giant hyssop, horsemint
(12)

Indians have eaten the seeds.

Agave parryi
Century plant
(24)

Indians would roast the large budding flower stalk until it was tender.

Agave spp.
Agaves

All species of *Agave* (with the exception below) have been recommended as possible food sources, to be tried with care. Methods of preparation would be as with either of the other entries on *Agave.*

One member of this genus, *A. lecheguilla*, should definitely be regarded as poisonous. It causes liver damage in some animals, along with photosensitization.
(16, 19, 24)

Agave utahensis
Utah aloe, agave

This has been much used as a food plant, along with its relatives. The center of

(16, 24)

the plant is pried out, and has been roasted for as long as three days. It can then be eaten as such or pounded flat into sheets and dried for storage. The seeds can be ground up for use as food. When roasted, the bulbous root is considered a delicacy by the Indians.

Agoseris aurantiaca
Mountain dandelion, false dandelion
(25)

Cooking the leaves makes them edible. One can also take the sap from the plant and harden it. This is said to be edible, or usable as a type of chewing gum.

Agropyron repens
Quackgrass, wheatgrass
(16)

In time of need, the long roots have been dried and ground into flour for breadmaking. As a substitute for coffee, they have been roasted and ground up. The seeds can be eaten.

Agropyron spp.
Wheatgrasses
(25)

The seeds of these plants are reported to be abundant. They can be ground up for flour or meal.

Agrostis spp.
Bentgrasses, redtop grasses
(25)

The seeds can be made into flour.

Alisma spp.
Water plantains
(25)

The bulbs are normally acrid; this is removed by cooking and drying them.

Allium spp.
Onion species

These can be prepared in any way that a domestic onion would be. The whole plant can be used.

Livestock losses have occurred where large amounts of members of this genus have been consumed. Humans should likewise reasonably limit the amount they eat at any given time.
(1, 8, 12, 15, 19, 24, 25)

Amanita caesarea
Orange amanita,
Caesar's mushroom
A. calyptroderma
A. citrina

Unfamiliar mushrooms should always be well cooked.

The deadliest mushrooms are *Amanitas*, so correct identification in this genus is very critical. Unless one is absolutely sure of identification, it is best to avoid *Amanitas* entirely.
(32, 35)

***Amanita muscaria**
Fly agaric

Specimens are parboiled in salt water until no more yellow scum floats to the surface. This probably contains the toxin, and is discarded.

This is one of the better known poisonous mushrooms, and is prized in Siberia for its hallucinogenic properties. Fly agaric should be prepared in only the above way.
(19, 32)

***Amanita rubescens**
Blushing venenarius
(35)

Unfamiliar mushrooms should always be well cooked.

Amanitopsis

Cook it well.

* See top entry of that genus.

vaginata
Sheathed amanitopsis
(35)

Amaranthus graecizans (A. blitoides)
Prostrate amaranth

The young plants are a popular green. The part above the root is cooked like spinach, and the taste is mild. By picking the mature plants, roughing them up and letting them dry out on something like a blanket, quantities of seeds can be obtained. These are edible raw or parched, or can be made into flour by grinding.

Nitrates have been shown to accumulate in members of this genus.
(16, 19, 24)

***Amaranthus hybridus**
Red amaranth, slim amaranth
***A. palmeri**
Palmer's amaranth
***A. retroflexus**
Green amaranth, redroot
***A. torreyi**
(15, 16, 19, 24, 25)

The young plants are popular raw, or when cooked by boiling in water until just tender. Drying and beating the mature plants yields quantities of small black seeds, which after winnowing can be eaten as is, parched or ground up for flour or meal.

Amelanchier spp.
Serviceberries
(1, 2, 15, 16, 24, 25)

Serviceberries have been much used by settlers and Indians in earlier days. Eaten raw, the seeds are bad tasting to some, but add to the flavor according to others. The berries are dried and eaten like raisins, or they can serve as the basis for jams, jellies, flavorings in stews, pies

* See top entry of that genus.

and so on.

Amphicarpa monoica
Wild peanut, hog peanut
(24)

The nuts of this plant are described as very pleasant eating. Squirrels and other rodents gather these nuts up for the winter, and a substantial part of the diet of Indians of the Missouri valley used to come from raiding these caches.

Amsinckia tessellata
Fiddleneck
(25)

The nutlet seeds can be ground into meal.

Anagallis arvensis
Scarlet pimpernel

It is used in salads or for greens in parts of Asia and Europe.

Occasional toxicity has been reported; moderation would be advised.
(19, 24)

Anemonella thalictroides
Rue anemone
(24)

The starchy tuberlike roots are edible when cooked. Medsger recommends that the plant be protected.

Anemopsis californica
Yerba mansa
(33)

The aromatic root is good raw. This plant is listed as having many medicinal uses.

Annona glabra
Pond apple
A. muricata
Soursop

The fruits are used in jams and jellies. Seminole Indians used the pond apple a lot.

A. squamosa
Sweetsop, sugar apple
(24)

Anthemis nobilis
Chamomile
(8)

A tea may be made from the flowers, fresh or dried.

Apios americana
Groundnut
(15)

The tuberlike parts of the root can be boiled in heavily salted water. They should be eaten hot, for when cold, they are reported to become tough and tasteless. The flavor of leftovers can be revived by roasting them for a while.

Apios tuberosa
Groundnut, wild bean
(2, 24)

The tubers may be eaten raw, but are reported to be better when boiled or roasted. Some say it tastes better than the potato. Medsger speculates that if man did not have the potato, he would probably have ended up cultivating these for food.

Apium graveolens
Wild celery, smallage

A tea can be made from the seeds. The whole plant is probably useful.

Apium tops have been reported to have contained toxic levels of nitrates.
(8, 19)

Apocynum cannabinum
Dogbane, Indian hemp

Seeds can be eaten after parching them, or they can be ground up into meal. The dried up sap has been used as chewing gum.

Various toxic principles have been isolated from this plant. Caution is advised in its use.

(19, 33)

Aquilegia spp.
Columbines
(33)

Spring leaves can be eaten after boiling them.

Arctium lappa
Great burdock
A. minus
Common burdock, clotbur
(2, 12, 15, 16, 24)

A. minus is popular enough to have been raised in Europe and Japan for food purposes. Young leaves and petioles can be eaten either raw or cooked. The inner, pithlike part of the stems is edible, and can be eaten raw, or parboiled if the flavor is too strong. Roots can be boiled until tender (younger roots are better since they are not so fibrous). Roasted and ground up, the roots serve as a substitute for tea or coffee.

Arctostaphylos spp.
Manzanitas
(12, 16, 24, 33)

The Indians made much use of all manzanitas. The berries were eaten raw or cooked, used in making jellies, or they were ground into meal for porridge.

Arisaema triphyllum
Jack-in-the-pulpit, Indian turnip

The roots have a stinging or burning quality which can be removed by drying them. Medsger recommends that the roots be first boiled, then dried, then ground into meal.

The irritant principle in the roots of *Arisaema* is held to be minute crystals of calcium oxalate. Oxalates are poisons, but probably in the amounts eaten here, insoluble calcium oxalate would have no effect on the body.
(19, 24)

Armillaria caligata
A. mellea

Two forms of *A. mellea* grow in North America, and they, as well as *A. cali-*

Honey mushroom *gata*, are described as choice.
A. ponderosa
A. zelleri
(24, 32, 35)

Artemisia abrotanum
Southernwood

Small quantities of chopped up leaves are used by some in salads. Dried, they can be used in pot-pourri.

Many species of *Artemisia* contain volatile oils which are toxic in larger amounts. Animals in grazing build up a tolerance to these toxins. Pending further research, this may or may not prove to be the case with humans as well. (8, 19)

***Artemisia biennis**
Sagebrush, biennial wormwood
(25)

The seeds can be eaten when cooked.

***Artemisia dracunculoides**
Russian tarragon
(8, 19)

The leaves can be used fresh or dried for flavoring.

***Artemisia dracunculus**
Sagebrush, tarragon
(8, 19, 25)

Seeds can be eaten if cooked, and the leaves serve as desired in flavorings.

***Artemisia tridentata**
Common sagebrush,

The plant is reported to be high in fat content. Grinding the seeds furnishes meal, and leaves can be steeped to make a tea which is rather strong.

* See top entry of that genus.

wormwood, mugwort (12, 13, 19)

***Artemisia vulgaris** Mugwort (8, 19) — Mugwort has been used to flavor beer, and can be used as a substitute for tea.

Arundinaria gigantea Large cane (24) — Young shoots can be cooked for a potherb; the starchy seeds were much used in earlier days and are said to be nearly equal to wheat in taste.

Asarum canadense Wild ginger (8, 24) — The aromatic rootstock can be dried and ground up for use as a substitute for ginger.

Asclepias incarnata Swamp milkweed — This is prepared as with *A. speciosa*. but does not rival it in taste.

All milkweeds should be regarded as poisonous plants. Resinoids are chiefly responsible for this action, although glycosides and an alkaloid have also been isolated from *Asclepias.* The broad-leaved species are generally less toxic, but potential gourmets should stick to the methods of preparation which are outlined.
(16, 19)

***Asclepias spp.** Milkweeds (19, 33) — Young plants can be parboiled twice for use as a potherb. Boiled roots are edible, and Sweet says that pods and stems of *Asclepias* have been eaten (presumably not raw). Caution is recommended in

* See top entry of that genus.

taking these steps. She also mentions that the hardened sap of the plant can be used for chewing gum.

***Asclepias speciosa**
Showy milkweed, pink milkweed, silkweed
(12, 16, 19)

This is referred to as an excellent food plant. Parboiling any of the parts of *Asclepias* helps to get rid of the milky juice which is bitter and probably toxic. Young shoots 4 to 8 inches long can be cooked like asparagus until tender. The youngest leaves on a plant which is more mature may be parboiled for a somewhat longer time than with spinach. Unopened flower buds are said to be delicious after boiling in many changes of water. The flower clusters contain enough sugar that they can be boiled down to make a thick syrup. Later on, the young boiled pods are good eating.

***Asclepias subverticillata (A. galioides)**
Poison milkweed
(16, 19)

This is one of the deadlier members of *Asclepias*. It is claimed that Indians destroyed the toxins by boiling and used this plant for greens. Indian children are reported to have eaten the flower buds.

***Asclepias syriaca**
Common milkweed
(15, 16, 19, 24, 25)

This can be tried as with *A. speciosa*.

***Asclepias tuberosa**
Butterfly weed
(16, 19, 24)

Try this as with *A. speciosa*. The roots can be boiled or baked for food.

* See top entry of that genus.

Asimina triloba
Papaw,
custard apple

The ripe fruit is sweet and is sometimes sold in markets.

This fruit provokes an allergic reaction in some people, in the form of a dermatitis or gastric symptoms.
(2, 15, 19, 24)

Asparagus officinalis
Asparagus

The shoots can be prepared in any of the ways that one would prepare common asparagus. The seeds of this plant can serve as a substitute for coffee, after roasting and grinding them. Stalks can be eaten raw or cooked.

Poisoning has been reported in cattle after eating nearly mature asparagus; the seeds are reported to be toxic at times.
(15, 16, 19, 24, 25)

Aster spp.
Aster species

The leaves of several of the *Asters* were boiled by various Indian tribes for food.

Some of the asters absorb selenium from the soil, and are therefore dangerous. For safety, asters should be avoided for food, unless it can be ascertained that they are free from selenium. See under *Salsola kali.*
(12, 19)

Astragalus caryocarpus
Ground plum,
buffalo pea
A. succulentus
(A. mexicanus)
Larger ground plum,
milkvetch,
buffalo pea
(16, 24)

Green seed pods can be eaten raw or cooked, or pickles can be made from them through cooking and spicing. Indians collected the seeds and cooked them for food.

Atriplex canescens Leaves can be cooked; Indians ground

Shadscale saltbush — the seeds up for meal.

This species is a facultative or secondary selenium absorber. Some of the other members of this genus also absorb selenium. See *Salsola kali.*
(19, 33)

***Atriplex spp.**
Saltbushes
(8, 16, 19)

Seeds or the leaves of any *Atriplex* are worth trying for food. The leaves are best when young and tend to have a high salt content which gives them added flavor.

Avena fatua
Wild oat
(16, 24)

Seeds of this plant were put into baskets with live coals and shaken by Indians. This would get rid of the hairs and parch the grains somewhat, after which they were ground up for meal.

Balsamorhiza hookeri
Hooker's balsamroot
(24)

The roots are edible raw, but are better when cooked.

Balsamorhiza sagittata
Arrowleaf balsamroot, balsamroot, bigroot
(1, 12, 24, 25, 33)

Seeds, young sprouts and roots are all edible raw or cooked. Olsen recommends that the roots be parboiled several times to remove a bitter taste. Young leaves are edible.

Baptisia tinctoria

The shoots are a good substitute for

* See top entry of that genus.

Wild indigo
(24)

asparagus.

Barbarea verna
Winter cress, scurvy grass
B. vulgaris
Common winter cress, yellow rocket
(2, 15, 19, 24)

These plants are popular with some Italians, who gather the young greens in early spring. They are good in salads or as a potherb. Older leaves can still be eaten by parboiling them to remove the bitterness. The flower clusters, even opened, can be substituted for broccoli.

Benzoin aestivale
Spicebush
(2, 24)

Historically, this plant has been much used for tea. It could be prepared from the leaves, twigs, bark or berries. Dried and crushed, the berries were used as a substitute for allspice.

Berberis spp.
Barberries, Oregon grapes
(16, 24, 32)

The berries can be eaten raw, cooked, or made into lemonade, jams or jellies.

Betula lenta
Black birch, sweet birch

Birches produce copious amounts of sap in the spring, which can either be drunk directly or boiled down into syrup. If disfigurement is no problem, the inner bark from the tree or the root is a valuable food. It is good eaten as is, or it can be cut up, dried and sealed in fruit jars for later use as tea. Leaves and twigs are also good for tea.

This and other members of the birch family contain methyl salicylate, or wintergreen oil. In minor quantities, this adds to the flavor and appeal of birches, but it is poisonous in larger quantities. The average lethal dose in adults is 30 milliliters.

(2, 15, 19, 24)

***Betula lutea** Yellow birch (19, 24)	The bark has a spicy flavor which is less pronounced than with *B. lenta.*
***Betula papyrifera** Paper birch, canoe birch (19, 24)	Indians made a tea from the leaves.
***Betula spp.** Birches (16, 19)	It would be worth trying to collect the sap from any birch tree.
Boisduvalia spp. Spikeprimrose species (33)	Seeds can be parched and ground up for flour.
Boletellus russellii (32)	Before eating the caps remove the tubes. The stalks of this mushroom are tough but may be worth trying if sufficient need exists.
Boletinus cavipes (32)	This mushroom is said to be choice.
Boletus aurantiacus & related species **B. chromapes**	Unfamiliar mushrooms should always be well cooked.

* See top entry of that genus.

B. chrysenteron
B. edulis
Steinpilz
B. frostii
B. mirabilis
B. pallidus
B. variipes
B. zelleri

Some boletes are poisonous. Those with red tube mouths are especially hazardous.
(19, 33)

Borago officinalis Borage	The peeled stems are good in salads, tasting like cucumber. A drink can be made out of leaves, stems or the flowering tops.

Borage has a high content of nitrate.
(8)

Brassica spp. Mustard species	When young, the plant can be eaten raw or cooked. As it matures, the leaves can be cooked for a potherb; they require fairly lengthy cooking. Mustard seeds are best known for their sharp flavor, and can be used for flavoring as the imagination dictates. Flowers and flower buds are palatable lightly cooked. *Brassica* pollen would be a good protein addition to the diet.

Mustards contain irritant oils and other toxins which demand some moderation in their consumption.
(2, 15, 16, 19, 24)

Brodiaea douglasii Wild hyacinth, cluster-lily, bluedicks (12)	The corm is good raw or better cooked. Tender seed pods are also a pleasant food.

Brodiaea spp. Wild hvacinths (12, 13, 24, 25)	Bulbs of any of these species are apparently edible raw or boiled.
Bumelia spp. (24)	The fruits of several species are edible.
Cactaceae Cactus family (16, 25)	Any cactus should be edible, especially in the younger stages. Coarse spines are usually burned off with a blowtorch. after which the flesh is roasted. Although irritating to the skin, the barely visible barbs of some species seem to produce no special ill effects when eaten. The fruits are pleasant. Seeds can be parched and ground into meal.
Cakile edentula American sea rocket (24)	The fleshy root has been pulverized and mixed with flour for making bread. The fleshy leaves can be used in salads or in cooking.
Calbovista subsculpta (32)	This is a popular mushroom.
Calendula officinalis Marigold (8)	Flowers can be dried for use in cooking or used fresh in salads.
Calla palustris Water arum, wild calla	Roots can be made into bread.

Like jack-in-the-pulpit, roots of this plant contain

minute crystals of calcium oxalate which must be detoxified before eating. This is presumably carried out as with *Arisaema triphyllum*.
(19, 24)

Callirhoe pedata Pimple mallow (24)	Indians eat the parsnip-like roots.
Calochortus nuttallii Sego lily, mariposa lily, star tulip, butterfly tulip (1, 12, 16, 24, 25)	There seems to be practically no way in which sego lilies cannot be prepared for food. The whole plant can be used raw, cooked, or whatever. The only danger would seem to be that this lovely plant might be wiped out through excessive use.
Calochortus spp. Mariposa lilies (16)	Any species should be edible.
Caltha leptosepala (C. rotundifolia) Marsh marigold, cowslip, meadowbright (12, 16)	Young leaves are good in salads or cooked, but they should be carefully washed before use. Older leaves are worth trying, and the boiled roots are supposed to be edible. The flower buds can be eaten raw.
Caltha palustris Marsh marigold, American cowslip	Greens can be boiled like spinach, which they surpass in flavor, according to some. Flower buds can be pickled and used in place of capers.

This plant has had a long history of toxicity. The volatile poisonous oil, protoanemonin, may not always be present in plant tissue, but can be driven off by sufficient heat when it is.

(19, 24)

Calvatia gigantea Giant puffball (32)	Some of these get as big as basketballs. Each one should be cut in half to verify that it is a puffball, and to observe the condition of the interior. The mushroom should be eaten only if it is white, never if the inside is slimy or powdery.
Calvatia spp. Puffballs (15, 32)	All puffballs are edible, and they range in taste from good to excellent. Make sure the interior is white, as with *Calvatia gigantea*.
Camassia leichtlinii (16)	This is presumably used as with *Camassia quamash*.
Camassia quamash (C. esculenta) Wild hyacinth, blue camass, camash (1, 12, 16, 24)	Indians commonly steamed and roasted the bulbs, but they are edible raw. They can be eaten apparently during any season, but without the identifying flowers there is the danger of confusing them with deadly *Zygadenus* bulbs.
Campanula rapunculoides Creeping bellflower, harebell **C. rapunculus** Rampion (8, 16)	Roots of either species are good after roasting, boiling or frying. Young leaves and shoots can be used in salads or for cooking.
Cannabis sativa	Marijuana seeds are edible. The shoots

Marijuana, hemp — are cooked as a substitute for asparagus in Belgium.

Most parts of the plant contain tetrahydrocannabinol, which gives the plant its narcotic reputation. Animals have died from eating it, and only the above plant parts should be tried for food.
(8, 16, 19)

Cantharellus aurantiacus
C. cibarius
Chanterelle
(15, 24, 32, 35)

Chanterelles are a gourmet item, and require slow cooking. They can be stewed, sautéed or dried.

Cantharellus cinnabarinus
C. clavatus
Pig's ears
(32, 35)

These are both described as choice.

Cantharellus floccosus
(32, 35)

This is an edible mushroom for some people; for others it is not edible. Caution is called for in its use.

Cantharellus infundibuliformis
C. minor
C. subalbidus
C. umbonatus
Grayling
(15, 32, 35)

Graylings are excellent after long cooking, and *C. subalbidus* is also choice. *C. infundibuliformis* is not recommended because of poor taste.

Capsella bursa-pastoris
(2, 16, 24, 25)

Young leaves are edible raw or cooked. They are delicious when blanched. Seeds have been ground into meal, or they can be used for flavoring other foods. Roots can be candied by boiling in syrup. Fresh or dried, they can be used in place of ginger. Pods are also useful for food.

Cardamine pennsylvanica Pennsylvania bitter cress **C. pratensis** Meadow cress, cuckoo flower **C. rotundifolia** Round-leaved watercress, American watercress **C. spp.** Bittercresses (24)	These top three are all good in salads; some of them are slightly bitter. Any bittercress is worth trying in salads.
Carex spp. Sedges (12)	Any sedge should be tried for food. The young shoots or bases of the leaves are tasty additions to the diet.
Carica papaya Papaya, papaw, custard apple (24)	The sweet fruit is often sold in markets.
Carnegiea gigantea Sahuaro (3)	Try the fruit.
Carya spp. Hickory trees (2, 15, 24)	Not all species of hickory nuts are edible, but are if they taste good. They have been sold in some markets, and can be prepared in any way desired,

such as combining in recipes, eating them raw, making soup out of them, or even pickling them. Sap of the trees can be used for food.

Cassia occidentalis
Coffee senna, coffeeweed

Roasted seeds are used as a substitute for coffee, even in Africa.

Coffee senna is a weakly toxic plant. Nothing is said of the toxin or the plant parts involved.
(4, 19, 24)

Castanea dentata
American chestnut, sweet chestnut, chestnut
C. nana
C. pumila
Chinquapin
(24)

These nuts are edible and choice; chestnuts are part of American history.

Castanopsis chrysophylla
Giant chinquapin
C. sempervirens
Bush chinquapin
(24)

The nuts are sweet and are related to chestnuts.

Catathelasma imperialis
(32)

This mushroom is not particularly good; it should be well cooked.

Ceanothus spp.
New Jersey teas,

Flowers or leaves can be dried and used for tea. This was done as far back as

buckbrushes, tobacco brushes, snowbrushes

the American Revolution. Indians ate the seeds. The berries of deer brush, *C. integerrimus*, are edible fresh or dried.

Some species of *Ceanothus* contain saponins, which are soaplike, poisonous glycosides. It is in the flowers and fruits, and was in fact used in earlier days in place of soap. If the plant part foams or lathers up, one should either leave it alone for food or remove the saponins. (See under *Yucca.*) (2, 8, 12, 13, 15, 16, 24)

Celtis crassifolia
Rough-leaved hackberry
C. mississippiensis
Southern hackberry
C. occidentalis
Hackberry, sugarberry
C. reticulata
Netleaf hackberry, western hackberry
(24)

The berries of each of these are edible, and some of them are highly esteemed.

Ceratonia siliqua
St. John's bread
(24)

The fleshy pods have been a favored food even in ancient times. They are eaten either fresh or dried.

Cercis spp.
Redbud, Judas tree
(33)

Flower buds can be added to salads or made into a kind of pickle.

Cereus giganteus
Giant cactus
(24)

The fruits are edible as is. They have also been made into preserves and a form of molasses.

Cetraria islandica
Iceland moss
(24)

The lichen has a bitter taste which is removed by boiling. It is then dried and pulverized for use as flour. This is an important foodstuff in Iceland.

Chaetochloa
Foxtail grass
(25)

Husks can be removed by parching the grain. It can be boiled for a mush or ground up for flour.

Chenopodium album
Lambsquarters, pigweed, common goosefoot

Young green parts of the plant have long been popular in salads. They can be served as a potherb after cooking a little longer than with spinach (they shrink a lot). Seeds are edible raw, or can be cooked or ground into flour.

Some members of *Chenopodium* have been shown to be weakly cyanogenetic, and *C. album* frequently contains toxic levels of nitrate. Neither of these toxins are cumulative ones, and have not detracted from the plant's appeal as food. A real danger in the case of *C. album* would seem to be oxalic acid, which may be present in plant tissue at levels of around 10% of the dry weight of the plant. See under *Oxalis violacea.*
(2, 15, 16, 19, 24, 25)

***Chenopodium ambrosioides**
Wormseed, wormseed goosefoot

This plant can be made into a strong aromatic tea.

This plant is the source of a poisonous terpene, ascaridole. Seeds of the plant are pressed for oil containing [illegible], and it is used in medicine as an anthelmintic.
([illegible], 19)

***Chenopodium**

Prepare as with *C. album.*

* See top entry of that genus.

berlandieri
Lambsquarters,
pitseed goosefoot
(2, 16, 19)

***Chenopodium bonus-henricus**
Good-King-Henry, wild spinach
(8, 19, 24)

Young shoots have been used in place of asparagus, and the greens were widely used as a potherb before spinach was introduced as a food.

***Chenopodium botrys**
Jerusalemoak goosefoot
(16, 19)

This plant is prepared for food as with *C. ambrosioides*, and the same dangers should be noted with regard to toxicity.

***Chenopodium capitatum**
Strawberry blite, blite goosefoot
(2, 16, 19)

Young greens are edible raw or cooked. They are still worth trying when older, after several parboilings. Fruit clusters can also be used as with the young greens.

***Chenopodium cornutum**
***C. fremontii**
Fremont goosefoot
***C. gigantospermum**
Mapleleaf goosefoot
***C. incanum**
***C. leptophyllum**
Slimleaf goosefoot
(16, 19, 24)

Try any of these as with *C. album.*

* See top entry of that genus.

***Chenopodium overi (Blitum hastatum)** (16, 19)	The fleshy fruit clusters should be edible.
***Chenopodium spp.** Goosefoot species (19, 24)	If necessary, the seeds of any goosefoot should be worth trying for food, with moderation and caution.
***Chenopodium urbicum** City goosefoot (19, 24)	This is another substitute for spinach.
Chimaphila umbellata Prince's pine, pipsissewa, waxflower (12)	The leaves or roots can be made into a beverage by boiling them.
Chiogenes hispidula Creeping snowberry, ivory plum (24)	An excellent drink can be prepared from the leaves and stems. The berries are an esteemed fruit.
Chlorogalum pomeridianum Amole, soap plant, wild potato (24)	The bulb is probably toxic raw, but it is nutritious when cooked.

* See top entry of that genus.

Chlorophyllum molybdites Cook it well and proceed with care.

This mushroom is regarded highly by those who are not made sick by it. Although apparently not lethal, it is poisonous to some people, and should be approached with care and respect.
(32)

Chondrus crispus
Irish moss, carrageen moss
(24)

This and other starchy seaweeds are useful for food. A jellylike substance can be leached out with boiling water, after which it is purified and used as a base for puddings and custards. The plant can be dried and used in making a beverage.

Chrysanthemum balsamita (C. majus)
Costmary, mint geranium
(8)

Young leaves give a hint of mint to salads.

Chrysothamnus spp.
Rabbitbrush species
(25)

Tender shoots at the tips of the plant can be eaten in the spring.

Cichorium intybus
Chicory, succory, blue-sailors
(1, 2, 8, 15, 16, 24, 25)

Young roots can be boiled like carrots. When older, they made a well known substitute for coffee. For this the root is peeled, roasted and ground up. When young, green parts of the plant are edible in salads or cooked. Growing older, they develop a bitter taste which can be removed partly or altogether by parboiling them or blanching them during growth.

Cirsium spp.
(Carduus spp.)
Thistle species

There seems to be little to the members of this genus, outside of the spines, which is not edible. Flowers, seeds, young leaves and the inner parts of the stems are all edible raw (a taste would certainly have to be cultivated for some of these). The roots, crown, and inner stems can all be cooked for food. Roasting also reduces a bitter quality to the seeds. A drink can be made from thistle leaves.

Canada thistle, *Cirsium arvense*, has been shown to contain toxic levels of nitrate, but this should be no deterrent to normal use of the plant for food.
(12, 13, 16, 19, 24, 25)

Cistus spp.
Sun roses
(8)

Leaves can be dried and used in flavorings.

Cladonia rangiferina
Reindeer moss
(24)

It can be eaten fresh, or made into a beverage described as stimulating.

Clarkia spp.
Clarkias
(33)

Seeds of this genus were collected by Indians. After drying and parching them, the seeds were ground up into meal.

Clavaria spp.
Coral mushrooms
(15, 32)

All coral mushrooms are reported to be more or less safe to eat, but some of them are too bitter to eat.

Clavariadelphus ligula
C. truncata
(32)

Unfamiliar mushrooms should always be well cooked.

Claytonia spp.
Spring beauties
(2, 12, 15, 16, 24, 25)

Apparently any spring beauty can be used for food, but it is often recommended that they be spared because of their rarity. If the need exists, the corms or roots can be eaten raw or cooked. The green parts can be eaten in some, if not all cases.

Cleome lutea
Yellow beeplant
C. serrulata
(C. inornata)
(Peritoma serrulatum)
Rocky Mountain beeplant, spiderflower, pink cleome, stinkweed

Indians made much use of *C. serrulata.* Seeds were pulverized for meal, and young green parts and the flowers of the plant were an important potherb. They would also boil the plant for a prolonged period of time, pull the plant parts out and continue boiling it down to a thick dark residue. This was used as paint, or left to dry for later food use.

C. serrulata has been shown to contain toxic amounts of nitrate.
(12, 16, 19)

Clitocybe alba
C. aurantiaca
Orange clitocybe
C. clavipes
C. infundibuliformis
C. multiceps
(32, 35)

Unfamiliar mushrooms should always be thoroughly cooked.

Clitopilus

These should be well cooked if unfam-

abortivus
C. prunulus
Prune mushroom
(35)

iliar. It is always worth checking mushrooms closely for worms and other small fauna.

Collybia acervata
C. butyracea
C. confluens
C. dryophila
Oak-loving mushroom
C. familia
C. radicata
Root mushroom
C. velutipes
Velvet collybia
(24, 32, 35)

Proceed with care with unfamiliar mushrooms and cook them well.

Commelina communis
Dayflower
(24)

This and other dayflowers can be used for food when cooked.

Condalia obovata
Purple haw, bluewood
(24)

The fruit is sweet and edible.

Coprinus atramentarius
Inky cap
C. comatus
Shaggy mane
C. micaceus
Glistening coprinus
(15, 24, 32, 35)

These mushrooms are all best when sautéed; the juice that cooks out of them is a flavorful addition.

Cornus canadensis Bunchberry, dwarf cornel, crackerberry **C. mas** Cornelian cherry **C. stolonifera (C. sericea) (C. instoloneus)** Red-osier dogwood, cornel, red willow (12, 16, 24, 38)	Various Indian tribes ate the fruits.
Cortinarius alboviolaceus **C. cinnamomeus** Cinnamon cortinarius **C. collinitus** **C. corrugatus** **C. violaceous** (35)	Unfamiliar mushrooms should be well cooked.
Corylus spp. Hazelnut species (2, 16, 24)	The nuts are a favored food.
Coryphantha arizonica Sour cactus **C. neo-mexicana** Spiny stars (3)	Try the fruits.
Crataegus spp. Hawthorns (12, 24, 25)	Twenty members of this genus have been recorded as being edible. Probably any hawthorn would be worth trying

for food. The fruits are eaten raw, or they can be dried or made into preserves.

Craterellus cantharellus **C. cornucopioides** Trumpet of death, horn of plenty (32)	These are considered choice mushrooms.
Crepis spp. Hawk's beards (25)	Leaves can be used in cooking.
Croton corymbulosus Chaparral tea, encinilla (24)	An excellent beverage can be made by boiling the flowering tops in water.
Cucurbita foetidissima Common gourd (33)	Indians have pulverized the seeds for food.
Cymopterus spp. Biscuitroots (16, 24, 25, 33)	Young green parts are edible raw or cooked. If they are too bitter, try parboiling them. Flowers are also tasty additions to the diet. Young roots are sweet when eaten raw, but older roots should be cooked. Roasted roots were pulverized by the Indians for meal, and the seeds of some species can also be used in making meal.

Cynoglossum grande
Grand hound's tongue
(33)

The cooked roots are edible.

Cyperus spp.
Flatsedges, chufas, earth almonds, nutgrasses
(16, 24)

These plants bear tubers which are historically famous for food. They can be utilized in just about any way imaginable, from eating them raw to candying them or making a beverage from them. They have to be peeled first. The inner base of the plant is edible raw.

Dalea terminalis
A species of indigo bush
(33)

Flowers were used by Indians for food. Because of their sweet taste, Hopi ate the roots for candy. Stems of the plant were boiled to make a tea good for various ailments.

Daucus carota
Wild carrot, Queen Anne's lace
D. pusillus
(16)

It is from *D. carota* that man set about to breed the domestic carrot. The roots are reported to be inedible raw, but dried and roasted, they make a substitute for coffee.

Dentaria spp.
Toothworts, pepperroots, crinkleroots
(2, 24)

Pepperroots are similar in taste to watercress, and can be eaten raw or cooked.

Dentinum repandum
D. umbilicatum
(32)

D. repandum has been ranked by some as among the most delicious of mushrooms.

Deschampsia spp.
Hairgrasses
(25)

Seeds can be used in making meal.

Descurainia spp.
(Sophia spp.)
Tansy mustards.
flixweeds

Young green parts of these plants can be steamed for about half an hour; they are then eaten or dried for future use. Parboiling helps to remove a bitter taste. Seeds were parched and ground by Indians into meal.

D. pinnata has caused poisoning in cattle which have eaten quantities of the plant over a period of time. Under proper treatment, the symptoms of poisoning eventually disappear. With moderate use for food, humans should have no difficulties.
(13, 16, 19, 33)

Diospyros kaki
Japanese persimmon
D. texana
Mexican persimmon, black persimmon
D. virginiana
Persimmon, date plum
(15, 24)

These persimmons are described as bearing luscious fruits. Leaves from *D. virginiana* are good fresh, or better dried in making a drink rich in vitamin C.

Disporum trachycarpum
Fairybells, mandarin
(12)

The berries are sweet and edible raw.

Dodecatheon hendersonii

Leaves and roots were roasted by California Indians. Other species of *Dode-*

Henderson shootingstar, American cowslip, birdbills (24)

catheon are probably edible.

Dracocephalum nuttallii
Dragonhead (25)

Try the seeds for mush or grind them into meal.

Echinocereus fendleri
Sitting cactus
E. munzii
Munz black-headed cactus
E. polyacanthus
Red-flowered torch cactus
E. rigidissimus
Rainbow cactus
E. triglochidiatus
Strawberry cactus
(3, 13)

The fruits of all of these are edible. Specific mention of edible seeds is given only in the case of *E. munzii*, but these and the fleshy portions are no doubt worth a try in any case.

Echinochloa crusgalli (Panicum crusgalli)
Barnyard grass, cockspur, watergrass

Harrington ground the chaff up along with the seeds into flour and reported good results with it.

Nitrate in toxic amounts has been found in this plant.
(16, 19, 25)

Elaeagnus argentea
Silverberry

Try the fruit.

(24)

Eleocharis palustris Spikerush (25)	Seeds should be tried in any of the accustomed ways.
Elymus spp. Wildrye species (13, 24, 25)	Probably any *Elymus* produces seeds which can be used for food.
Ephedra spp. Mormon teas, Brigham teas, jointfirs (16, 25)	Greenish stems from any species should be worth trying in tea making. They have been used for this when fresh, dried or roasted. Seeds are roasted for food.
Epilobium angustifolium (Chamaenerion angustifolium) Fireweed, willow herb **E. latifolium** Broad-leaved fireweed, red willow-weed **E. spp.** Willow-weeds (2, 12, 16, 24)	When young and not bitter, the green stalks and flower stalks are edible raw or cooked. Older leaves are still useful in making a tealike drink. The inner pithlike part of grown plants is flavorful and desirable, but difficult to gather in quantity. Some of the smaller species are used in the same ways.
Equisetum arvense Field horsetail, scouring rush **E. fluviatile**	The base of the plant, its underground parts and the peeled stems are reported to have been used by Indians for food. The cone-bearing shoots can be cooked like asparagus; parboiling them during

E. hyemale
Western scouringrush
E. laevigatum
Smooth horsetail

this process may help to get rid of undesirable substances. The green stems of some species are pulverized for use as a mush or in cooking.

Livestock have been poisoned through eating horsetails. There are various chemicals in *Equisetum* which may contribute to this, but the most troublesome factor has been found to be a thiaminase which reduces or removes the thiamine content of the body. No indication is given that the toxin or toxins in *Equisetum* are cumulative. Pending further information, horsetails are probably acceptable as food, with caution.
(16, 19, 25, 33)

Eriodictyon californicum
Yerba santa
(33)

Leaves can be boiled for tea.

Eriogonum spp.
Eriogonums, umbrella plants
(24, 25)

Seeds of these plants can be prepared by grinding them for flour, etc. The leaves can be used in cooking. Young inflated stems of *E. inflatum* are tender when eaten raw, and are made by some people into pickles.

Erodium cicutarium
Storksbill, filaree, pinkets, alfileria
(12, 16)

One can use the young greens raw or in cooking.

Erythronium americanum
Yellow adder's

These are recommended by authors only as emergency foods; their delicate beauty is too easily destroyed. Leaves or

tongue, fawn lily **E. grandiflorum (E. parviflorum)** Dogtooth violet, snow lily, fawn lily, adder's tongue (12, 16, 24, 25)	corms are edible raw or cooked, but when raw, are said by some to be emetic.
Fagus grandifolia (F. americana) American beech (2, 24)	Beech leaves can be cooked. The nuts are good, but they are rather small. If an emergency exists, the inner bark can be dried and pulverized for food.
Ferocactus viridescens Green hedgehog cactus **F. wislizeni** Barrel cactus (3)	Fruits of the green hedgehog cactus are edible. Fleshy parts of the barrel cactus can be candied.
Festuca spp. Fescue grasses (25)	The seeds are a food item.
Fistulina hepatica Beefsteak mushroom (15, 24, 32)	This is recommended broiled or stewed.
Flammulina flavida F. velutipes Winter mushroom (32, 35)	Only caps of the winter mushroom should be cooked for food (discard the gelatinous part of the cap and the stalk).

Foeniculum dulce Florence fennel **F. officinale** Fennel (8)	The flavor of stems of *F. dulce* is said to be delicate when cooked, and fennel is useful fresh or dried in flavorings.
Fragaria spp. Strawberries (1, 2, 12, 15, 16, 24)	Besides strawberries, the leaves can be made into a tea described as excellent tasting.
Frasera speciosa (Swertia radiata) Green gentian, deertongue, monument plant giant frasera (12)	Indians are reported to have eaten the roots.
Fraxinus spp. Ash species (16)	Sap of an ash can be used for food.
Fritillaria atropurpurea Leopard lily, tiger lily, purple-spot fritillaria **F. pudica** Yellow fritillary, yellowbell **F. spp.** Fritillarys	The corms of these plants are valued items, and are a regular part of the diet of Indians and Eskimos. The green seed pods, raw or cooked, are described as delicious. The corms of *F. pudica*, at least, are surrounded by great numbers of ricelike bulblets, which taste like rice when cooked.

Until more is known, one fritillary, *F. meleagris*, should be avoided. It is reported to contain a heart-depressant alkaloid, and has caused poisonings.

(12, 19, 25)

Fuscoboletinus ochraceoroseus **F. spectabilis** (32)	*F. ochraceoroseus* leaves a bitter aftertaste with some people.
Galera tenera (35)	Cook unfamiliar mushrooms well.
Galinsoga ciliata **(G. aristulata)** **G. parviflora** Littleflower quickweed, quickweed, galinsoga (16)	Young green parts of quickweed are much used as a potherb in other countries. The taste is described as bland.
Gaultheria humifusa Western wintergreen, creeping wintergreen, checkerberry, teaberry **G. procumbens** Wintergreen, checkerberry, partridgeberry, mountain tea (2, 15, 16, 24)	One can just chew on the leaves, or use them fresh or dried in making tea. The fruits are spicy and edible fresh, or are commonly used in pies and preserves.
Gaultheria shallon Western wintergreen, sallal (2, 16, 24)	Indians placed great value on this fruit. They would dry it for winter use.

Gaylussacia spp.
Huckleberries
(15, 24)

These fruits are excellent in flavor.

Geum rivale
Avens,
water avens
(12, 16, 24)

One can cut the roots into pieces and boil them for a drink that looks like chocolate.

Gleditsia triacanthos
Honey locust,
honey shucks
(24)

There is a sweet pulp between and around the seeds which is regarded highly. Indians are said to have used the pods and seeds for food.

Glyceria spp.
Mannagrasses
(25)

The seeds are flavorful and are used in any of the ordinary ways.

Glycyrrhiza glabra
Common licorice
G. lepidota
Wild licorice,
American licorice

The roots have a sweet taste and can be eaten as is.

When quite young, *G. lepidota* is poisonous.
(8, 12, 24)

Godetia biloba
Lobed godetia
(33)

Indians parched and pulverized the seeds for winter use.

Gomphidius rutilus
G. vinicolor

Treat these as one would any unfamiliar mushroom.

(32)

Graminae The seeds of any grass should be tried
Grass family for food if necessary.

One should exercise caution that all black masses are excluded from the grain gathered. Poisonous fungi will often grow where the grain is, and they are lethal if eaten in sufficient quantity.
(16, 19, 25)

Grindelia squarrosa Gumweed, resinweed, tarweed — Indians chewed the leaves and used them also in making a drink.

If selenium is present in the soil, gumweed will absorb it. One should therefore avoid this plant for food, unless selenium is definitely shown to be absent. See *Salsola kali*.
(12, 19)

Gymnocladus dioica Kentucky coffee tree, American coffee bean, chicot — Pioneers used the seeds for coffee, by roasting and grinding them as with coffee beans. The seeds are supposed to have been roasted and eaten as such by Indians.

Various parts of this tree are poisonous, including the seeds. Roasting apparently detoxifies them.
(2, 19, 24)

Gyrodon merulioides — Cook unfamiliar mushrooms well.
(32)

Gyroporus castaneus — These mushrooms are choice.

G. cyanescens
(32)

Habenaria dilatata
White bog orchid, bog candle, scent bottle
(12)

Orchids should be used in emergencies only, for they are generally rare. Indians and Eskimos eat the tubers after boiling them for half an hour or so.

Hedysarum boreale
Northern sweetvetch
H. mackenzii
Licorice root
H. occidentale
Western sweetvetch, sweetbroom
(12, 24)

The roots of these plants have been used raw for food.

Helianthus annuus (H. aridus)
Sunflower
(12, 15, 16, 24)

Sunflower seeds of the wild variety are too small to crack individually, but some Indian tribes parched them and ate them along with the hulls. Some writers recommend cracking the hulls with a food grinder set to the right size. The hulls float to the top when the results are dumped into water. The seeds can also be pulverized and boiled until their oil floats to the surface, which is useful in cooking. Sunflower seeds can furthermore be roasted and ground up for use as a coffeelike substitute. Yellow flower petals can be used for food, and young flower heads have been boiled and eaten like brussels sprouts.

Helianthus maximiliani Maximilian sunflower (12)	Tubers can be eaten raw, boiled or roasted.
Helianthus petiolaris Prarie sunflower (16)	Try this as with *H. annuus.*
Helianthus spp. Sunflowers (16)	The seeds of all species can undoubtedly be used as with those of *H. annuus.*
Helianthus tuberosa Jerusalem artichoke (2, 12, 16, 24, 25)	The tubers of this plant can be prepared as with those of *H. maximiliani.*
Helvella californica **H. caroliniana** **H. esculenta** Beefsteak morel **H. gigas** Snow mushroom **H. infula**	Some of these are choice mushrooms. They should be well cooked.

Three of these are poisonous to some people and edible to others. *H. esculenta* is described as an excellent mushroom, but some people have died from eating it. (The other two are *H. californica* and *H. infula.*) It is best to start out in such cases, if at all, with very small amounts. Individual reactions can then be noted without excessive danger.
(32)

This section is devoted to the identification of some of the more common plants and classes of plants found in the listing. For a broader treatment of plants and their identification, there are several excellent texts out. Any of these are highly recommended for specific identifications:

1. Craighead, J. J., F. C. Craighead Jr. & R. J. Davis:
A Field Guide to Rocky Mountain Wildflowers, Houghton Mifflin Co., Boston, Mass., 1963.
2. Harrington, H. D:
Edible Native Plants of The Rocky Mountains, University of New Mexico Press, Albuquerque, New Mexico, 1967.
3. Kirk, Donald:
Edible Wild Plants of The Western United States, Naturegraph Co., Healdsburg, Calif., 1970.
4. Medsger, O. P:
Edible Wild Plants, Macmillan Co., New York, 1966. (The eastern United States is covered in this.)

One desiring to go further and identify plants in a systematic fashion can do so by obtaining one of the keyed works covering plants in his area. In step-by-step elimination of plant characteristics, one systematically narrows possibilities down to the plant in question. Most universities have herbariums, as well, where personnel are traditionally happy to help in identifications.

Plate 1

Agastache urticifolia
Giant hyssop

This is an aromatic perennial growing from 2-6 feet high. It is found in damp soil to about 8000 feet. Its locale of growth is from Montana to Colorado and from there westward, but other members of the mint family occur throughout North America. The size of this plant and its dense spikes of flowers serve to distinguish it from other mints.

Members of the mint family are easily identifiable by their square stems. The stems of this plant are hollow, and branch several from each root. Leaves are opposite, ovate to deltoid, toothed, stemmed, and are 1½-3″ long.

Asclepias speciosa
Showy milkweed

This perennial grows from 2-6 feet tall. It prefers moist areas and waste places, at medium elevations and below. It can be found from Minnesota to Texas and from there westward. Nearly 100 species of *Asclepias* occur through most of North America.

Individual flowers are up to 1″ wide. They are found in dense umbels 2-3″ across, and are whitish to pale pink as shown. The petals are thick, and have a soft frosted quality to them. They bloom in June and July.

Leaves are opposite. They are thick and leathery, and are oblong to lance-shaped. The veins in the leaves are characteristic. Seeds mature in September, in rough, conspicuous pods from 3-5″ long. The numerous seeds are flat and tufted with long, light hairs to catch the wind. The plant contains a milky sap.

Cirsium undulatum
Wavyleaf thistle

This is a perennial, growing 1-4 feet tall. Its flowers bloom from June to September. As a genus, *Cirsium's* most distinctive characteristic is the predominance of spines on the herbage and bracts. The flowers are purple to rose-colored, but can vary to whitish. They are disc flowers only. The seeds are small, dry and smooth. *Cirsium* is widespread through North America, and about 150 species grow in other parts of the world.

Epilobium angustifolium
Fireweed

These perennials grow from 2-8 feet tall, in moist soils and areas that have been burned over or disturbed. It is widely distributed in the United States, and is most common between 6000 feet and timberline. It is most often confused with other members of the evening primrose family, but is distinguished by its large size. Other *Epilobiums*, about 100 of them, occur in most non-tropical regions of the world.

The flowers of *E. angustifolium* occur in terminal racemes that bloom from July to September. They can be as much as 1″ across. Buds, flowers and mature seed pods can appear on a plant at the same time. The pods are from 2-4″ long, and later split into 4 parts to release seeds tufted to catch the wind. The leaves are lance-shaped, and from 2-6″ long.

Agastache urticifolia

Asclepias speciosa

Cirsium undulatum

Epilobium angustifolium

Plate 2

Equisetum hyemale
Western scouringrush

Horsetails are common in moist situations. They are fairly distinctive in that they have no leaves as normally understood, and no flowers. *E. hyemale* forms hollow, pencil-like stems which are jointed and unbranched. The leaves are very small and tooth-like. They attach to each other, and surround the stems at the nodes.

Fragaria vesca
Wild strawberry

Various wild strawberries grow throughout the world; this one is found through North America. It is a perennial which grows a few inches high, and is found in moist soil at most elevations. This might be mistaken for barren strawberry (*Waldsteinia idahoensis*), which has yellow flowers and lacks the characteristic runners.

Strawberries are probably too familiar to need description. They spread by runners. The flowers, as illustrated, are white, with 5 petals seperate from each other. There are 20 stamens to a flower, and each flower is about ¾″ wide. The leaves are soft and thin, and consist of toothed leaflets as shown.

Lactuca serriola
Prickly lettuce

This is an annual that grows in dry areas. It grows from 1-5 feet tall, and is found over most of North America.

The flowers are yellow, but often fade to a bluish tint. They are from ½-¾″ long. The seeds are achenes, with dandelion-like hairs to catch the wind.

The leaves are alternate, toothed, and project themselves partially around the stem. The center rib of each leaf on the underside has spines on it; these spines are found also on lower parts of the stem. As shown in the photo, leaves can vary in shape from deeply cleft to lobeless. The variety of *Lactuca* with lobeless leaves is regarded as a cross between the cleft type of *Lactuca* and cultivated lettuce. The sap is milky.

Equisetum hyemale

Fragaria vesca

Lactuca serriola

Plate 3

Eriogonum heracleoides
Umbrella plant

This is a member of the buckwheat family. It is a perennial, usually under a foot high, in dry or open locations up to 9000 feet. It grows from Nevada to Colorado, up to Alberta and from there westward. The many members of this genus might be hard to distinguish from each other, but since *Eriogonums* as a genus are edible, this is not critical. There are some 150 *Eriogonums* located through North America.

The flowers of this plant are cream-colored, or often rose-colored in the earlier stages. They bloom from May to July, in compound umbels as shown.

The plant arises from a woody base, and tends to be mat forming. The leaves are linear to lance-shaped, pubescent, and are green above and whitish below. Flowering stems usually have leaf-like appendages at about mid-point.

Potentilla fruticosa
Shrubby cinquefoil

This member of the rose family is a spreading, branched shrub from 1-5 feet high. It grows up to 9000 feet in the mountains, and prefers damp or rocky areas. It is distributed through most of the world.

The flowers bloom from June to August, and average about an inch across. The leaves are 1/3-1″ long, are pinnately compound, leathery, sessile, and range from elliptic to linear. They are occasionally evergreen. The bark tends to shred off.

Ribes aureum
Golden currant

This can be found in moist situations in the mountains, often growing in clumps. It grows from 3-10 feet tall, and is thornless, with glabrous herbage. It is located mostly from Saskatchewan to Mexico and from there westward. Other gooseberries and currants are located through North America.

The flowers are bright yellow, and are borne in racemes out of leaf axils. They bloom from April to June. The berries are many-seeded, about ¼″ across, range from yellow to red to black, and are available mostly from August to September. The leaves are alternate, palmate, with 3-5 lobes, measure from 1-1½″ wide, and are long-petioled.

Rorippa nasturtium-aquaticum
Watercress

Watercress is an evergreen member of the mustard family. It grows in ponds, slow-moving streams and mud, and is found throughout North America. Other cresses may be mistaken for watercress. If the plant in question grows in water and has white flowers, it will be prostrate if watercress and erect if not.

The flowers are white and bloom from May to September, in racemes as shown. The pods are from ½-1″ long. The leaves have a turnip-like taste and like a radish, are hot. The roots are white and threadlike.

Potentilla fruticosa

Eriogonum heracleoides

Ribes aureum

Rorippa nasturtium-aquaticum

Plate 4

Rosa woodsii
Wild rose

This plant grows from 3-10 feet high, in damp soils. It is common in thick patches, commonly mingling with other plants along streams. It can be found up to 7000 feet, from Saskatchewan to New Mexico and from there westward. Other roses are common throughout North America.

The flowers of wild rose bloom from May to July, and average about 2" in diameter. The hips contain several seeds. They turn red in the fall, and adhere to the plant through winter. The leaves are toothed and pinnately compound.

Rubus parviflora
Thimbleberry

This delightful shrub grows up to 10,000 feet in valleys and mountains. It grows from 2-6 feet high, along streams and in moist shady spots, from Ontario to New Mexico and from there westward. Other species of *Rubus*, which number in the hundreds, are located mostly in the north temperate zone.

The flowers bloom from May to July, and can reach 2" across. The fruits have a soft, velvety texture, are bright red, and surpass, in the writer's opinion, most other berries in flavor.

The leaves, as illustrated, are large, usually about 6" across, and are palmately lobed, toothed, and lighter green on the underside. The plant is thornless.

When it rains in the woods, one can often tell if there are thimbleberries in the area. The plant gives off a lovely fragrance when rained on.

Rumex patiencia
Patience dock

This belongs to the buckwheat family, and is a perennial. It grows from 2-6 feet tall, and is found in many parts of North America. Various species of *Rumex* are common to North America, and some are found in most parts of the world.

Seeds are borne in conspicuous panicles as shown; each seed is sectioned into 3 parts by brownish-red, winglike valves. The leaves tend to be broadly lanceolate, and are up to a foot long. The stems are often red-tinged.

Sedum stenopetalum
Yellow stonecrop

This is a fleshy perennial which grows from 4-8" tall. It prefers rocky locations, and is most common from 6000-9000 feet. It occurs from Saskatchewan to New Mexico and from there westward. Sedums can often be confused with others of the genus, but this should not matter since all are edible. Other Sedums occur throughout North America.

The yellow flowers are sometimes tinged with purple, and are borne in dense terminal clusters as shown. They bloom from June to August. The leaves are alternate, ¼-½" long, thick and fleshy, and range from green to reddish-brown.

Rosa woodsii

Rubus parviflora

Rumex patiencia

Sedum stenopetalum

Plate 5

Helianthus annuus
Sunflower

Sunflowers are annuals that can grow around 11 feet high under cultivation, but the wild variety is very much shorter. It is very common in dry soils, below 7000 feet, over most of the United States. A number of plants could be mistaken for the sunflower; the way to make sure is to examine the seed. Viewed from the top, the seed of a true sunflower will appear 4-angled. (This may require a magnifying lens.) About 60 other species of *Helianthus* are native to the Americas.

The flowers are usually 4-5″ wide, and terminate the stems. They bloom from July to September. Stems and leaves of the plant are quite coarse and hairy. The leaves are ovate, from 2-10″ long, and are toothed. Lower leaves of the plant are commonly heart-shaped.

Heracleum lanatum
Cow parsnip

These are perennials which grow from 3-10 feet high. They grow in damp places, from sea level up to 9000 feet, over most of North America. They can be distinguished from other members of the parsley family by the leaves, which are surpassingly large.

The flowers are borne in compound umbels as shown, and are usually white, but can vary to a mild pink. The umbels are up to a foot across, and bloom from May to July.

Leaves measure up to 1½ feet, or are occasionally larger. They are coarsely toothed, ovate, and in compounds of 3. The plant in the foreground is a fireweed.

Iliamna rivularis
Mountain hollyhock

This plant grows from 3-6 feet high in damp situations, up to 9000 feet. It is a perennial herb tending to grow in large clumps, and is found from Alberta to Colorado and from there westward to Nevada and British Columbia. It might be confused with other members of the mallow family, but is generally much taller.

The flowers are borne in racemes which terminate the stems. They are about 2″ across, and can vary from rose-purple to white. Many yellow stamens grow from the center of each flower. They start to bloom at the bottom of the racemes, and the flowering period is from June to August. Leaves are maple-like in shape, are toothed, and are from 2-6″ wide.

Linum lewisii
Blue flax

This little perennial grows from 1-2 feet high, and is found in dry locations between 4500 and 9000 feet. Most commonly, it is found from Saskatchewan to New Mexico and from there westward. Harebell (*Campanula rotundifolia*), is most often mistaken for this plant, but its flowers are bell-shaped and the petals are united.

The flowers are usually sky blue and saucer-shaped, or occasionally white. They are usually borne one at a time, and range from ½-1″ across. Each flower has 5 petals, 5 sepals and 5 stamens. The petals are seperate. They bloom from June to August.

Leaves are alternate, simple, very narrow, and are ½-1″ long. Several stems arise from a woody crown; they are very slender, fibrous, and hard to break.

Helianthus annuus

Heracleum lanatum

Iliamna rivularis

Linum lewisii

Populus tremuloides
Quaking aspen

These trees are tall and slender, and are widespread in mountains above 6000 feet. They often form large stands in moist locations. They are located throughout Canada and the northern and western United States.

The tree gets its name from the leaves, which in a mild breeze will flutter. The leaves are toothed and long-stemmed, and average up to 2″ long. The bark is smooth and white.

Prunus virginiana
Chokecherry

Many of our domestic fruits, such as plums, apricots and peaches belong to the genus *Prunus*. Chokecherry is a native shrub or tree that can grow to 30 feet tall, and prefers moist situations. It usually grows below 8000 feet, and is found throughout North America.

The flowers, as shown, are borne in racemes up to 4″ long. They are white with yellow stamens, and are pleasantly fragrant. Individual flowers are usually under ½″, and bloom in May and June. The fruits are red, turning black when ripe. Their flavor is pleasant, but quite astringent, which is the reason for the plant's common name. Each cherry contains a single hard smooth stone, about 3/8″ in diameter.

The leaves are shiny, toothed along the margins, oval, and stemmed. They range from 1-4″ long, and are among the first leaves to turn color in autumn. The branches and twigs are brown, thornless, and hairless. It is a member of the rose family.

Tragopogon dubius
Goatsbeard

These biennials are quite distinctive. They grow from 1-4 feet tall, and are found in many situations, from dry to moist. They are common throughout the United States. *T. pratensis* and *T. porrifolius* are very similar to this plant in appearance. The flowers of *T. porrifolius* are purple, however.

The flowers are surpassed in length by bracts, which can measure up to 3″ when the plant fruits. Seeds, as illustrated, are borne in conspicuous heads like dandelions, which measure 2-3″ in diameter. Leaves are linear-lanceolate, and are from 4-12″ long. The sap is milky.

Populus tremuloides

Prunus virginiana

Tragopogon dubius

Plate 7

Amelanchier alnifolia
Serviceberry

This member of the rose family is a shrub or small tree which grows up to about 7500 feet in the mountains. It can be found from Ontario to New Mexico and from there westward. It grows up to 20 feet tall, but is usually much shorter. About 25 members of the genus grow in the north temperate zone.

The flowers occur in short racemes and are quite fragrant. They grow up to 1" wide, and bloom in May and June. The berries are purple when ripe, and are from ¼-½" wide. The berries ripen in late summer and are pomes.

The leaves are from 1-2" long, and are usually rounded at the base and toothed toward the apex. They range from oval to nearly round. The plant is thornless.

Ceanothus velutinus
Snowbrush

This member of the buckthorn family is an aromatic, spreading shrub which is common in the mountains in dnese patches. It grows from 3-8 feet high. It prefers moist soils, grows up to 8500 feet, and is found from South Dakota to Colorado and from there westward. Most other members of the genus grow in California.

The flowers are small and in showy panicles. The petals, sepals and stamens of each flower number 5. The leaves are quite thick and leathery, tend to curl, and are evergreen. They are ovate, alternate, glossy green on top, dull green beneath, and are finely toothed along the margins. Each leaf has 3 prominent veins.

Glycyrrhiza lepidota
American licorice

This member of the pea family is a perennial that grows from 1-3 feet high. It prefers moist soils, and grows below 7000 feet. It can be found from Saskatchewan to Mexico and from there westward.

The flowers bloom from May to August. They are about ½" long, and occur in dense racemes. The pods are brown, about ½" long, and as shown, are covered with hooks to help in their distribution.

The leaflets are about 1" long. They are arranged in pinnately compound fashion, in groups of 11-19.

Grindelia squarrosa
Gumweed

This biennial is easy to identify by the sticky substance which covers the flower bracts. It grows from 1-2½ feet tall, in dry soils, and is found from Minnesota to Texas and from there westward. It grows to 8000 feet.

The flowers are about 1¼" across, and are surrounded by the series of narrow bracts. They bloom from July to August. The leaves are more or less toothed.

Amelanchier alnifolia

Ceanothus velutinus

Glycyrrhiza lepidota

Grindelia squarrosa

Plate 8

Osmorhiza chilensis
Sweet cicely

This plant is found in damp, shaded areas, and grows from 1-3 feet tall. It is much branched, and not aromatic as *O. occidentalis* is. The flowers are borne in compound umbels, and are an inconspicuous greenish-white or pinkish. The seeds are linear-oblong, and are barbed to help in their distribution.

Various species of *Osmorhiza* are found throughout North America. They are perennials with thickened roots. The flowers are borne in umbels, with the narrow seeds tapering at one or both ends. They are members of the parsley family.

Osmorhiza occidentalis
Sweetanice

This prefers cool, moist locations, and grows up to 3 feet high. It is easy to identify from the leaves, which when crushed, smell heavily of licorice. The flowers are yellowish-green, and bloom in May and June.

Rhus radicans
Poison ivy

In gathering plants for food, one of the most painfully educational experiences can occur from coming into contact with poison ivy, oak or sumac. It is hoped the reader will be familiar enough with these plants to avoid the results of contact. Most widespread among these is poison ivy, which ranges from southern Canada and the United States into Mexico.

Poison ivy occurs as a shrub or vine 2-6 feet high. It prefers moist, shady locations. The leaves are glossy on top, glabrous, but tend to vary in size and shape. The only constant feature to look for are the 3 terminal leaflets. In the fall, the leaves turn a bright red. The flowers are white and inconspicuous, bloom in axillary panicles from May to July. An aid to identification in winter are the berries, which remain on the plant in winter and early spring. They are white, glabrous, and about 3/8″ in diameter.

The plant is toxic irregardless of season. Even burning it can be dangerous- contact with smoke from the fire can be worse than touching the plant itself. Shown with poison ivy in the photo is *Equisetum hyemale.*

Sambucus caerulea
Blueberry elder

This is a bush or a small tree which grows from 6-18 feet tall. It grows in mountains in moist locations, and ranges from Montana to Utah and from there westward. *Sorbus scopulina* (mountain ash) is easy to mistake for this plant, but its leaves are alternate.

The flowers are clustered in showy panicles as shown, and give rise to small berries which turn from green to purple when ripe. The berries measure about ¼″ across.

The leaves are pinnately compound, with 5-9 leaflets. They are opposite, lanceolate, finely toothed, about 2-4″ long, and have a distinctive odor when crushed. The stems are hollow, straight and jointed. It is a member of the honeysuckle family.

Osmorhiza chilensis

Osmorhiza occidentalis

Rhus radicans

Sambucus caerulea

Hemerocallis flava
Daylily
(16)

The tuberlike roots are probably edible. Flower buds can be prepared for food as with *H. fulva.*

Hemerocallis fulva
Daylily,
Common orange daylily
(16)

Young shoots are often used raw or cooked, for food in Japan. Tubers, when young, are supposed to be excellent when eaten raw or cooked for around forty minutes in salty water. Peeled, older roots can also be cooked up into an acceptable dish. Flowers of the plant are picked for food either before the flower is open or after flowering. They are pleasant when cooked in various ways, and are similar in this respect to green beans. They can also be dried for later use.

Heracleum lanatum
(H. maximum)
Cow parsnip,
masterwort,
Hercules-parsnip

Tender shoots can be added to salads, or cooked: They may have a rank odor, which can be reduced by parboiling them. The fibrous peeling should be removed from older shoots before cooking them. Mature stems can be prepared for food in much the same way as the shoots. Roots taste like rutabega when cooked. The base of the plant is said to be useful as a substitute for salt; for this it is presumably dried and pulverized.

One writer has suggested the roots may be poisonous.
(12, 16, 33)

Hericium caput-ursi
H. coralloides
H. erinaceus
H. weirii
(32)

Reports are sketchy on the edibility of *H. weirii*, so caution is advised. It is important to cook *H. erinaceus* slowly to tenderize it; otherwise it may cook up too tough.

Heteromeles arbutifolia (Photinia arbutifolia) Christmas berry, toyon berry, California holly (24, 33)	Berries have been boiled, baked, parched and ground up into meal, and used in making drinks.
Heuchera micrantha Alum root (33)	Indians would boil or steam the leaves for food, or they would steam and dry them for future use.
Hippuris tetraphylla **H. vulgaris** Marestail (16)	The parts of these plants above the root are supposed to be tender in any stage of growth, and can be used in cooking. Alaskan Indians in particular made use of them– during winter the plant occasionally protrudes above frozen waters.
Holodiscus discolor Mountainspray, oceanspray, rock-spirea, creambush **H. dumosus** Bush rock spiraea, bush oceanspray (1, 12)	The small dry berries were used as food by the Indians.
Hydnum caput-ursi Bear's head **H. coralloides** Coral fungus	Bear's head and coral fungus are excellent mushrooms. *H. imbricatum* should first be sampled raw. If mild in taste, it should be edible; if the taste is strong or peculiar, do not eat it.

H. imbricatum
(24, 32)

Hydrophyllum appendiculatum
Waterleaf
H. capitatum
Ballhead waterleaf, woolen-breeches, pussyfoot
H. spp.
Waterleaf species
H. virginianum
Virginia waterleaf
(12, 24, 25)

The roots and young shoots of probably any waterleaf are worth trying for food. Shoots of the first and last entries are prized in the raw state, and with *H. capitatum* at least, the whole plant can be cooked up for food.

Hygrophorus cantharellus
H. chlorophanus
H. conicus
H. miniatus
Vermilion mushroom
H. pratensis
Meadow hygrophorus
H. puniceus
H. russula
(32, 35)

Cook them well if they are unfamiliar. *H. russula* is said to be one of the best of mushrooms.

Hypericum spp.
St. Johnsworts, klamath weeds

Indians dried the whole plant and pulverized it into a meal, which they used in cooking. They also ate the fresh leaves.

This genus contains at least six species, worldwide, which are poisonous. The toxin is a pigment, hypericin, which causes photosensitivity in the skin of animals who eat it. Upon exposure to light, such skin will form lesions, seep, itch, or fall off in more severe cases. There is no

complete recovery from the aftereffects of this poisoning, so it is nothing to take a chance on. The plant responsible for most poisonings in the United States has been *H. perforatum*, and *H. concinnum* is also suspect. In any case, the toxic pigment is confined to small glandular dots on the foliage and flower petals. Where a *Hypericum* has such dots, it would be logically safe to eat the plant after the parts containing these are stripped off.
(19, 33)

Hypholoma appendiculatum **H. incertum** **H. perplexum** Red hypholoma **H. sublateritium** (24, 35)	Cook it well if it is an unfamiliar mushroom.
Hyssopus officinalis Hyssop (8, 24)	Dried flowers can be used for making tea. With moderation, sprigs of the plant can serve as a flavoring.
Ilex cassine Dahoon holly **I. glabra** Appalachian tea, evergreen winterberry **I. paraguayensis** Paraguay tea, maté, yerba **I. verticillata** Black alder, winterberry **I. vomitoria** Cassena, yaupon, Indian black drink (24)	The leaves of these plants are popular for the drinks that can be made from them. Indians were very fond of cassena.

Iliamna rivularis (Sphaeralcea rivularis) Mountain hollyhock, maplemallow, globemallow, (24)	Indians used stems of some of the hollyhocks for chewing gum.
Impatiens biflora Spotted touch-me-not, jewel weed (24)	Young stems, available around spring, are said to be edible.
Ipomoea leptophylla Bush morning glory **I. pandurata** Wild potato vine, mecha-meck, man-of-the-earth (16, 24)	These roots sometimes yield up a surprising amount of food. The bush morning glory is said to have even larger roots than those of *I. pandurata*, which can weigh thirty or more pounds. The roots can be roasted or boiled. Harrington found the younger roots of the bush morning glory less woody and very acceptable raw. They can be dried for storage.
Juglans spp. Walnut species (2, 15, 24)	The nut of any *Juglans* should be tried for food— raw, in cooking, or any other way.
Juniperus communis Mountain common juniper **J. occidentalis (Sabina occidentalis)**	The fruits of *J. occidentalis* and other species of western junipers were sometimes pulverized and made into cakes by Indians. The berries are used by some people in flavorings.

California juniper, western juniper (8, 13, 24)	
Juniperus spp. Junipers (16)	Any juniper should be considered, if necessary, as a source of food.
Juniperus scopulorum (Sabina scopulorum) Rocky Mountain juniper, red cedar (16)	Many uses of the berries of this juniper are described. Indians used to eat them raw; under the resinous taste they are sweet. They can be boiled or roasted, or dried for storage. Roasted and ground, they are a substitute for coffee; or they can serve as a tea. The inner bark can be eaten, if one's need is great enough. Young shoots can also be used in making tea.
Labiatae Mint family (25)	Seeds or leaves can be used in making soothing teas.
Laccaria laccata Waxy mushroom **L. ochropurpurea** (32, 35)	Cook *L. ochropurpurea* slowly. *L. laccata* is not recommended because of difficulties in identifying it properly.
Lactarius aurantiacus **L. camphoratus** **L. corrugis** **L. deceptivus** **L. deliciosus** Delicious lactarius **L. indigo**	Some of these are choice mushrooms, while others are marginal. Some of them require slow cooking for proper results.

L. lignyotus
Sooty lactarius
L. piperatus
L. sanguifluus
L. subdulcis
Sweetish lactarius
L. theiogalus
Sulfur-milk lactarius
L. thyinos
L. torminosus
L. vellereus
Fleecy lactarius
L. volemus
Orange-brown lactarius

Kingsbury gives one classification of mushrooms according to the types of poisoning they produce. In this, *Lactarius* mushrooms are mentioned categorically as causing enteritis. People still eat them, however, so the possibility of this danger cannot be too overwhelming. Avoid any *Lactarius* where the juice is milky and turns yellow in air, or where wounds stain purplelike in color.
(15, 19, 24, 32, 35)

Lactuca canadensis
Canada lettuce
L. ludoviciana
L. pulchella
Blue-flowered lettuce, chicory lettuce
L. serriola
(L. scariola)
(L. integrata)
Prickly lettuce, wild lettuce

Young green parts are acceptable in salads, or they can be used as potherbs. These shoots are somewhat bitter, but this can be reduced by parboiling them, or protecting the growing plants from light. Roots of *L. pulchella* yield up a gum used as chewing gum by the Indians.

Cattle which have fed on large quantities of the young stages of *L. scariola* have suffered from massive lung damage. Even so, it is no doubt safe to eat any of the above species in moderate quantities. The toxic principle is apparently inactivated when the plant at any stage of growth is dried. (12, 16, 19, 24)

Laportea canadensis (Urticastrum divaricatum)
Wood nettle
(24)

This is recorded as being edible. It can no doubt be prepared for food as the stinging nettle is, where younger green parts are used as a potherb.

Lateiporus sulphureus
Sulfur shelf
(32)

The only edible part of this mushroom is the tender part along the rim, which is cut off two or less inches back from the edge.

Ledum groenlandicum
Labrador tea

Leaves can be dried to make a tea, or people chew on them.

Caution is advised with this plant. Members of its genus and family (*Ericaceae*) are common poisoners of livestock. owing their toxicity to resinoid principles. The poisoning appears to be noncumulative, but is responsible for pathological conditions which may be hard to recover from. (2, 24)

Lemaireocereus thurberi
Organ pipe cactus
(3)

This is another of the cacti whose fruit is edible.

Lemna spp.
Duckweeds
(16)

These are aquatic plants that can yield up much salad material.

Lentinus

Pick young specimens only and use them

lepideus
Scaly lentinus
(32, 35)

when fresh. Slow cooking of this mushroom is required. If it is too tough to eat, a good soup can be made out of it.

Lepidium spp.
Pepperweeds
(24)

Seeds can be processed as desired into a condiment.

Lepidium virginicum
Peppergrass
(24)

The leaves are occasionally used in salads or as a garnish.

Lepiota americana
L. naucina
L. procera
Parasol mushroom
(35)

Use them thoroughly cooked.

Lepista irina
L. nuda
(32)

These are both described as excellent mushrooms, but *L. irina* disagrees with some people.

Leucoagaricus naucinus
L. procerus
Parasol mushroom
L. rachodes
(32)

The latter two species are gourmet items. *L. naucinus* is not recommended because there are difficulties involved in properly identifying it, and it does not agree with everyone who eats it.

Levisticum officinale
Lovage
(8)

Leaves are useful fresh or dried in flavorings.

Lewisia rediviva
Bitterroot,
rockrose,
bitterroot lewisia
(12, 16, 24)

It is best to collect roots when the plant is in flower. The outer layer of the roots is most easily peeled off then. This contains most of an objectionable bitter taste. The roots can be boiled or parboiled until they soften, and are good, but still have a touch of bitterness. They can also be baked or pulverized to form a meal.

Liatris punctata
Gayfeather,
blazingstar,
button snakeroot
(12)

Try the corms.

Lilium spp.
Lily species
(16, 24)

Several different lilies were used by Indians as one would use potatoes; they are raised for food in Japan. In the United States, they are rather rare in the wild state at least, and deserve protection.

Limacella lenticularis
(32)

The *fischeri* variety of this mushroom is probably edible, but only probably. Some people have recommended the *lenticularis* variety as food.

Linum lewisii
Blue flax,
wild flax,
prarie flax

The seeds are said to be pleasant in taste, and are used in any way that grain usually is.

This species has been shown to have cyanogenetic potential, which is no deterrent for use as food, as long as such use is moderate.
(19, 24)

Liquidambar styraciflua
Sweet gum, red gum, bilsted
(24)

The hardened sap or resin has been used as chewing gum.

Lithospermum incisum
(L. angustifolium)
Gromwell, puccoon, Indianpaint
(12)

Indians in the West cooked the roots for food.

Lomatium spp.
Desert parsleys, biscuitroots

The roots used to be a major article of trade among the Indians. They were eaten raw, cooked, or ground up into meal which was shaped into cakes. Seeds of some species are useful when ground into meal. No *Lomatium* has been reported to be poisonous, but a person would have to get used to the taste of some of the species.

These should be carefully identified. Some members of the parsley family are lethally toxic.
(12, 24, 25)

Lonicera ciliosa
Honeysuckle, orange honeysuckle
L. involucrata
Twinberry
L. utahensis
Red twinberry
(12, 24)

The berries are edible, and are recorded as ranging in taste from poor to good.

Lupinus spp.

The poorer people of days past boiled

Lupines

the seeds to remove a bitter taste. In early spring, Indians would steam the leaves and flowers for food.

Several lupines are poisonous. A number of alkaloids have been isolated from them, and an unidentified toxic factor has been investigated in western Australia. At least some of the alkaloids involved (which are concentrated mostly in the seeds) are noncumulative in sheep. They are not harmed by eating forage over a period of one day, that in one feeding would kill them. The non-alkaloidal factor found in Australia causes icterus in livestock, and the liver damage involved is cumulative. With different lupines at different stages of growth, different symptoms are suffered by different animals. Even though some of them are all right for food, lupines are best avoided where a person has a choice. Identification within the genus is difficult.
(19, 24, 33)

Lycium carolinianum
Boxthorn
L. fremontii
L. halimifolium
Matrimony vine
L. pallidum
Wolfberry, pale wolfberry, desert thorn, box thorn, rabbit thorn, tomatilla
L. spp.
Matrimony vines, wolfberries
L. torreyi
Torrey wolfberry

Lycium berries have been commonly used when fresh. They can be cooked into a soup or sauce which can be eaten or dried for later use. Practically all of the species found in New Mexico were used by the Indians. *L. halimifolium* has been used as a potherb.

L. halimifolium has been listed as being poisonous to animals.
(16, 24)

Lycoperdon spp.
Puffball mushrooms
(15, 24, 32)

These, as with the *Calvatias*, are edible and popular. Any hint of yellow on the inside indicates over-age, and should be culled out; one such mushroom can spoil the whole pot.

Lygodesmia grandiflora
Rushpink, skeleton plant
(12)

The leaves were boiled with meat by Indians.

Lygodesmia juncea
Skeleton weed

Indians of the Missouri valley gathered stems of this weed and cut them so that the juice would exude out. When it hardened, they would use it for chewing gum. The plant is often infested with round galls which contain gum, and this was used for the same purpose.

The plant is recorded as possessing toxic concentrations of nitrate.
(19, 24)

Lysichitum americanum
Yellow skunkcabbage, American skunkcabbage

Young leaves are edible after repeated parboilings. Indians used the roots by roasting, drying and pulverizing them into flour.

The stinging properties of the plant are caused by calcium oxalate. This can be inactivated by heating or drying.
(12)

Madia sativa
Chile tarweed
(24)

Seeds can be used for pinole. Medsger suggests that some of the tarweeds native to the United States could probably be used in the same way.

Magnolia virginiana
Sweet bay, laurel magnolia (24)

Leaves are recorded as being good in flavorings.

Mahonia repens (Berberis repens)
Holly grape, Oregon grape, creeping barberry, mountain holly (12, 25)

Berries have been eaten raw, or they can be dried or made into jellies or juice.

Malus spp.
Wild apples

Wild or crab apples can be eaten fresh or green, if one is up to it. They are excellent in apple recipes.

Eating green apples can produce abdominal pain. Animals that have eaten large quantities of fallen apples stagger about and show other symptoms of poisoning, and deaths have occurred. Apple seeds contain a cyanogenetic glycoside, and at least one man has died from eating too many of them at one time (Apple seeds cannot help but contain nutritious material, though. Poisoning will not occur if by eating small amounts of them, the body has sufficient time to metabolize away the minute quantities of cyanide). (15, 19, 24)

Malva neglecta (M. rotundifolia)
Mallow, cheeses, cheese weed, round-leaved mallow
M. parviflora

M. neglecta is a versatile plant which can be nicely prepared for food in many and various ways. Young green parts of the plant are good in salads, and are mild tasting. They have a mucilaginous juice which recommends them for soup making. Young plants can be dried for use as

M. spp. **M. sylvestris** High mallow **M. verticillata** Whorled mallow, curled mallow (16, 24)	a tea. The green fruits are a popular nibble, and they can be used at any stage of growth. (The clinging sepals do not have to be removed.) These fruits are good in cooking and can be pickled. Seeds of *M. sylvestris* are also mentioned as a food item. Probably any *Malva* is worth trying in this or any of the other ways.
Marasmius oreades Fairy-ring mushroom (15, 24, 32, 35)	This is a popular mushroom, and is excellent in stews or sautéed. It can be dried for storage.
Marrubium vulgare Horehound (1, 8, 24)	Leaves are useful in flavorings, and are the basis for horehound candy.
Martynia louisiana (Proboscoidea lousianica) Devilsclaw, unicorn plant, double-claw (16, 24)	When young, the tender pods are good in pickle recipes. They might be worth trying in soups. The pods of some of the devilsclaws growing in the southwestern states were a source of food for the Indians.
Medeola virginiana Indian cucumber root (24)	In taste and smell, the tuberlike root is like a cucumber.

Medicago lupulina Black medic, nonesuch (24)	The seeds of this plant were greatly relished by Southern California Indians.
Melissa officinalis Lemon balm (8, 24)	Leaves can be used to flavor salads or other dishes. A refreshing beverage can also be made from them.
Mentha spp. Mint species (2, 8, 12, 24, 25)	As well as serving to make comforting teas, mint leaves are useful in flavorings. The seeds are edible, or can be steeped along with the leaves for tea.
Mentzelia spp. Blazingstars (12, 25)	Blazingstar seeds have been parched and pulverized into a meal useful in various ways.
Mesembryanthemum crystallinum Sea fig, ice plant **M. edule** Hottentot fig (24)	The stems and leaves can be used in salads.
Mimulus spp. Monkeyflowers (1, 12, 33)	Leaves and stems are crisp additions to salads. If they are slightly bitter, as with *M. guttatus*, parboiling or picking them when younger may help.
Mitchella repens Partridge berry, twinberry	Berries of this plant hang on throughout the winter, and are pleasant eating.

(2, 24)

Monarda spp.
Beebalm species, horsemints
(2, 8, 12, 24)

Leaves are useful raw, for making tea, in flavorings, and as a potherb. *Monarda* flowers can be dried for use in cooking.

Monolepis nuttalliana
Nuttall monolepis, poverty weed, patata
M. pusilla
(16)

Younger green parts of the plants make an excellent potherb. Roots can be cooked up for food as well (the older ones get a little tough). The seeds are edible.

Montia perfoliata (Claytonia perfoliata) (Limnia perfoliata)
Miner's lettuce, Indian lettuce

The leaves and stems are good raw or cooked, and are rich in vitamin C. Flowers can be eaten.

Nitrate has been found in miner's lettuce, which should be no deterrent to normal use.
(1, 2, 16, 19, 24, 25)

Montia chamissoi
Water montia
(16)

This is worth trying for food, probably as with *M. perfoliata.*

Morchella angusticeps
Black morel
M. crassipes
Thick-footed morel

Morels are choice and coveted mushrooms. They are probably best sautéed.

M. esculenta Morel **M. hybrida** Half-free morel (15, 24, 32)	
Morus alba White mulberry **M. microphylla** **M. nigra** Black mulberry **M. rubra** Red mulberry (2, 15, 24)	Mulberry fruits are generally sweet and choice, and are very acceptable fresh or in recipes. Shoots of *M. alba* can be consumed raw or cooked.
Mycena galericulata (35)	Thoroughly cook unfamiliar mushrooms.
Myrica asplenifolia (Comptonia peregrina) Sweet fern (2, 24)	Leaves have been used, fresh or dried, in making tea. The fruit is also edible.
Myrica gale Sweet gale, bog myrtle (8, 24)	Leaves can be used in flavoring recipes, as well as the berries.
Myrrhis odorata Sweet cicely (8)	Roots or leaves can be used in salads for their background flavor. Boiled, the roots have been used as food.

Naematoloma capnoides **N. fasciculare** **N. sublateritium** Brick cap (32)	Only mild tasting specimens of *N. fasciculare* should be used for food. Avoid them if they taste bitter. This species of mushroom is probably edible for some people, and for others mildly poisonous.
Naucoria semiorbicularis (35)	Use it well cooked, as with any unfamiliar mushroom.
Navarretia spp. Skunkweeds (33)	Seeds were parched and pulverized for food.
Nelumbo lutea American lotus, water chinquapin, American nelumbo, water beans (24)	This really lovely plant is rare, and should be spared for that reason. However, probably any part of the plant is edible. Indians prized the seeds, which can be boiled or roasted. (There is a hard shell on each seed which should be cracked so the water can soften the contents faster. If they are to be roasted, it should be done while the seeds are fresh and their shells are still soft.) The tuberlike roots resemble sweet potatoes in flavor.
Neomammilaria applanata Nipple cactus (3)	Fruits are edible.
Nepeta cataria Catnip **N. hederacea**	Catnip leaves are good for making tea, and poorer people in England have often used *N. hederacea* for the same purpose.

(Glechoma hederacea)
Ground ivy, gill-over-the-ground, creeping Charlie

Horses are poisoned by eating large amounts of ground ivy, either fresh or when dried. Its toxicity is probably due to the presence of volatile oils. These should not cause trouble in moderate quantities.
(8, 15, 19, 24)

Nuphar advena
N. polysepalum
(N. luteum)
(Nymphaea polysepalum)
Rocky Mountain cowlily, Indian pond lily, spatterdock, yellow waterlily
(12, 16, 24, 33)

Cowlily roots are rather massive. Indians would boil or bake them to remove the outer covering, and use them in various ways for food. The inner root can be dried and pulverized for flour. If there is a strong taste to the flour, it can be removed through washing and careful decanting of water off the flour. Muskrats store the roots for food, and Indians used to raid these caches, as could anyone in need. Seeds are also useful for food. Various ways can be used to remove the hard shell around each seed. They can be parched and pulverized into meal, or when sufficiently parched, eaten like popcorn.

Nyssa aquatica
(N. uniflora)
Large tupelo, wild olive
N. sylvatica
(N. multiflora)
Sour gum, black gum, pepperidge, tupelo
(24)

Fruits of either species are used in preserves.

Oenanthe sarmentosa
Water parsley
(24)

The starchy tubers are good boiled. Be absolutely certain the tubers are not those of the lethal *Cicuta douglasii.*

Oenothera spp.
Evening primroses
(2, 12, 16, 24, 25)

Any member of this genus should be tried, if necessary, for food. They might be tried as *O. strigosa* is, which has been cultivated in Europe for food purposes. Young greens can be used in salads, or parboiled long enough to remove a bitter taste, if it is too strong. Young roots and the inner part of older roots can be boiled or parboiled. Seeds of many *Oenotheras* can be parched for food or ground into meal; and unripe pods of many, when cooked, make acceptable eating.

Omphalia campanella
(35)

Cook unfamiliar mushrooms well.

Onoclea sensibilis
Sensitive fern

Infant fronds can be picked as they are unfolding and used as a potherb.

Horses have suffered poisoning from eating hay containing this fern. The sumptoms were delayed or were absent where horses received exercise or were able to eat grain. Parallels between animals like horses and humans are tentative at best, but if one is inclined to try plant parts other than what is mentioned, it would be best in this case to avoid the mature plant.
(19, 24)

Onoclea struthiopteris (Matteuccia

Roots were boiled or roasted by Abnaki Indians for food.

struthiopteris)
Ostrich fern
(24)

Opuntia spp.
Prickly pears, chollas
(2, 3, 12, 16, 24, 25)

As with *Cactaceae*, the pads can be prepared and eaten after disposing of the thorns or barbs. People have eaten this flesh boiled or fried, and it has even been dried for storage. The fruits are good, and seeds are useful in cooking, after parching and grinding them into meal.

Orchidaceae
Orchid family
(12)

Bulbs of all North American orchids are recorded as being edible. But they deserve protection since they are rare.

Origanum spp.
Marjorams
(8, 24)

Leaves are useful in making drinks or in culinary flavorings.

Ornithogalum umbellatum
Star-of-Bethlehem

Cooked bulbs are pleasant eating.

The bulbs have caused poisoning in livestock, and this is probably from an unidentified alkaloid. Animals have not been found to be poisoned unless the bulbs are brought to the surface through something like plowing or frost heaving. Other parts of the plant have had no poisonous effects on grazing animals.
(19, 24)

Orobanche spp.
Broomrapes
(24, 25)

Some *Orobanches* have roots which can be cooked in various ways like potatoes.

Orogenia linearifolia
Indian potato, great basin orogenia, turkey peas, snowdrops
(1, 12, 16, 25)

The tasty tubers are good raw, or can be fried, baked or boiled. The outer peeling can be taken off if its taste is objectionably bitter.

Orontium aquaticum
Golden club, taw-kee
(24)

The bulblike roots can be cooked for food. After repeated boilings, seeds were also eaten.

Oryzopsis asperifolia
Mountain ricegrass
O. hymenoides
(Eriocoma hymenoides)
(E. cuspidata)
Indian ricegrass, Indian millet, Indian mountain-rice
(16, 24, 25)

Seeds can be used in probably any way that other grains such as wheat are. One analysis of the seeds found them to contain about 18 to 20% starch and 6% of various sugars.

Osmorhiza spp.
Sweet cicelys, sweetroots
(16, 24)

The roots and fruits of any of these should be tried for food. The roots in particular are recorded as being heavy with a sweet licorice or aniselike flavor, and where this is too strong, the roots and seeds can be dried and pulverized for use as a food flavoring. Where the taste is weak enough, roots can be boiled until soft and the rind peeled off. The taste is close to that of parsnips. While still fleshy, fruits are pleasant nibbles.

Oxalis acetosella
True wood sorrel
O. violacea
Violet wood sorrel

Leaves have a pleasing acid taste and are used in salads.

All species of *Oxalis* should be regarded as potentially dangerous. The toxins involved, oxalic acid or one of its soluble salts, are hazardous substances to risk eating. In moderate quantities, they pass through the body with no ill effects, apparently. At some greater concentration which the consumer may have no way of recognizing, complex changes occur in his body, poisoning him. The foremost pathological change is a lowering of blood calcium, and fouling of kidneys and other parts of the system with the resultant crystals of calcium oxalate. Perhaps it would be safest with suspect plants to test what has been gathered for soluble oxalate content. (This is done by crushing a representative sample in water, and testing the filtrate with calcium ion. The amount of precipitate which is formed will be proportional to the amount of soluble oxalates present.) One thing that will make any such plant safe to eat is the addition to it when eaten of sufficient calcium ion. The oxalates then react preferentially with this calcium before or in the stomach, and pass through harmlessly. One may simply choose to eat moderately of oxalate containing plants, as many people have done for a long time, and nothing may happen. (Yet people who have thought everything was all right have found out later in their lives in some hospital that they had kidney stones, composed usually of calcium oxalate. Surprisingly, some common foods have been shown to contain soluble oxalates, among them beet tops, spinach and rhubarb.) (8, 19, 24)

Pachystima myrsinites
Mountain lover,
Oregon boxwood,
myrtle boxleaf

Berries can be used in cooking.

mountain-hedge
(25)

Panaeolus retirugis
(35)

Unfamiliar mushrooms should be well cooked.

Panax quinquefolius
Ginseng
(8)

Roots are used, and have well known medicinal properties.

Panus strigosus
(35)

These mushrooms should be used only when young.

Papaver rhoeas
Corn poppy,
red poppy
P. somniferum
Opium poppy

Poppy seeds are much used in some parts of the world for food. They are pleasantly nutty like sesame seeds when used in place of them, and can be pressed for their edible oil.

Both of these and other poppies contain toxic alkaloids, which should limit their use as food to the seeds.
(8, 19)

Passiflora incarnata
Maypop,
passion flower,
passion vine
P. lutea
Yellow passion flower
(24)

The fruits of these are food items esteemed by some, and can be used in preserves or beverages.

Paxillus atrotomentosus

Thoroughly cook them.

P. involutus
(32, 35)

Pedicularis lanata (12)	The yellow roots taste like carrots and can be prepared for food like them.
Pelargonium capitatum Oak-leaf geranium (8)	Leaves can be dried for use in cooking.
Pellaea ornithopus Bird's-foot fern, tea fern, bird's-foot cliff brake (24)	Leaves and stems of this evergreen fern can be steeped to make a pleasant tea.
Peltandra virginica Green arrow arum, Virginia tuckahoe (24)	It is reported that the boiled spadix and berries were regarded as luxuries by the Indians. They roasted the large bulbous roots for a day or two in pits, then dried them and ground them up for meal.
Peltiphyllum peltatum (Saxifraga peltata) Indian rhubarb (24)	The thick leafstalks are a gourmet item to the Indians. They can be peeled and eaten as is, or cooked.
Peniocereus greggi Sweet potato cactus (3)	Fruits are a good item.

Peraphyllum ramosissimum
Squaw apple
(16)

The fruits can be eaten raw if necessary (they are sour or somewhat bitter, depending on their age). They have been used in jellies and in cooking.

Perideridia bolanderi
Bolander's yampa
P. gairdneri
(Carum gairdneri)
(Atenia gairdneri)
Yampa,
wild caraway,
squawroot
(12, 16, 24, 33)

Yampa roots were highly regarded by Indians; to some of them it was their favorite root food. There is apparently no way in which the roots cannot be prepared into tasty food, if one does not want to just eat them raw. About all that one cannot do with them, apparently, is to store them away fresh like potatoes. They store easily when dry.

Persia borbonia
Sweet bay,
red bay, laurel,
Isabella wood
(24)

Dried bay leaves are much used in flavorings, and are important in Creole cookery.

Phacelia ramosissima
Branching phacelia
(12)

Indians in California used the plant for greens.

Phaseolus acutifolius
Tepary bean
P. polystachyus
Wild bean,
bean vine
P. retusus
Prarie bean
(24)

These beans are very good when cooked. *P. polystachyus*, and probably the others, can be dried for storage.

Pholiota caperata
P. discolor
P. precox
P. squarrosa
Scaly pholiota
P. squarrosa-adiposa
P. squarrosoides
(32, 35)

Only younger scaly pholiotas should be eaten. When preparing *P. squarrosa-adiposa* for food, wipe the slime off the caps and discard the stalks. Smith regards *P. squarrosoides* as the best food mushroom in the group.

Phragmites communis
Reed, common reed grass
(24, 25)

Reeds are high in sugar content. When the stalks are mechanically injured, or through insect attack, sap oozes out and hardens into gum. This the Indians made into balls to eat. Another use of the sugar content was made by drying the stalks and grinding them up, and sifting the flour out. This flour swells and browns near fire like marshmallows do, and becomes a taffeelike confection. Seeds and roots are also edible.

Phylloporus rhodoxanthus
(32)

Use only the caps of this mushroom.

Physalis spp.
Ground-cherry species, husk-tomatoes

Yellow-flowered members of this genus are particularly recommended as worth trying for food, but any species deserves a try. The ripe fruits are good raw, cooked, or dried for future use.

The foliage and unripe berries of at least one *Physalis* (*P. peruviana*) are poisonous.
(2, 4, 12, 15, 16, 24, 25)

Phytolacca americana

Poke shoots, while their leaves are still unrolled, are excellent eating. They can

(P. decandra)
Pokeweed, poke, scoke, garget, pigeonberry

be cooked like spinach or parboiled (which some writers report to be necessary for safety in eating). The shoots can also be fried or pickled. Some people have made pies out of the berries.

Pokeweeds contain saponins, soluble oxalates, and other unidentified poisonous principles, which appear to be present in every part of the plant including perhaps the shoots. The roots are especially poisonous. Berries are regarded as less toxic, but may have caused deaths in humans. If there is a burning sensation in the mouth after eating poke, do not proceed further. Unless the poisoning is fatal, humans are said to recover in about one day. There is always a question about oxalates, however. See under *Oxalis violacea.*
(4, 15, 19, 24)

Picea mariana
Black spruce
P. rubra
Red spruce
(24)

Young leaves and twigs have been used to make "spruce beer."

Pinaceae
Pine family
(Abies, Cedrus, Larix, Picea, Pinus, Pseudotsuga and **Tsuga spp.**

Raw or boiled, the inner bark of members of the pine family can be eaten.

Cattle have suffered birth abnormalities after eating fallen needles of *P. ponderosa*, and there have been fatalities among them from eating needles of *P. taeda* (loblolly pine).
(2, 19)

Pinus lambertiana
Sugar pine
(2, 13, 24, 33)

Sap of the sugar pine is sweet, as the name suggests. As the sugar dissolves, the resin left behind might serve as chewing gum.

Pinus sabiniana
Digger pine
(33)

Green cones have a soft center which the Indians roasted for about twenty minutes. This would turn syrupy, and was a much appreciated food.

Pinus spp.
Pine trees

While some species of pine are better for food than others, any species should be worth a try if personal need demands. Boiled pine needles are said to give a drink containing vitamin C; if the beverage is too strong, it could be diluted, or they could be steeped for perhaps a shorter time. Some pines yield nuts that are small, but they are all edible. For prolonged storage, Indians would roast the pinenuts. They also cracked the nuts, winnowed out what shells they could and pulverized them to make a meal useful in various ways. In emergency conditions, the inner bark can be stripped or scraped off. It has been eaten raw, or can be cooked or dried and pulverized into meal. Tender pine needle shoots have been eaten raw.

As with *Pinaceae*, pine needles in some cases are potentially toxic.
(2, 16, 19, 24, 33)

Pinus strobus
White pine
(24)

The pitch of this tree has been used as a questionable substitute for chewing gum.

Plantago lanceolata
Buckhorn plantain, narrow-leaved plantain

The leaves of any of these are suitable potherbs. Young leaves of *P. major* are used in salads, and its dried leaves can be used in making tea.

P. major
Common plantain,
Indian wheat
P. rugelli
(2, 16, 24)

Pleurotis ostreatus
Oyster mushroom
P. sapidus
Oyster mushroom
P. ulmarius
Elm mushroom
(15, 24, 32, 35)

With the oyster mushroom, the base of the funnel shaped cap is tough and should be cut away before using it for food. These are all recorded as being excellent mushrooms.

Pluteus admirabilis
P. cervinus
Pluteus
P. magnus
(24, 32, 35)

P. cervinus is an excellent mushroom, but reports on the edibility of *P. magnus* are sketchy.

Poa spp.
Bluegrasses
(25)

Gather the seeds for food. As with many grasses, they do not remain on the plant for long, so the season for picking them is short.

Podophyllum peltatum
Mayapple,
mandrake, wild lemon

The ripe fruits are delicious, but should be eaten in moderation.

The fruit is the least toxic part of the plant, which owes its toxic nature to podophyllin, a mixture of 16 or more physiologically active substances. No other parts of the plant should be used for food. Deaths or poisonings have occurred through ingestion of young shoots, foliage, and the roots in particular.

(2, 15, 19, 24)

Polygonatum biflorum True Solomon's-seal (24)

Young plants cooked like asparagus are described as an excellent dish. Indians and others in need have helped sustain themselves on the starchy root.

Polygonum bistorta Bistort (8)

Roots have been a source of starch when no other options were available. Leaves have been used in making Easter Puddings.

Polygonum bistortoides (Bistorta bistortoides) (B. linearifolia) American bistort, western bistort, knotweed, snakeweed (12, 16)

Young greens are useful in cooking. The roots are pleasant eating, raw or cooked, but older ones that are fibrous would be better cooked.

Polygonum cuspidatum Japanese knotweed, Japanese fleeceflower (15, 16)

When taken before the leaves start unfolding, young shoots can be cooked like asparagus, and have a tart taste. After peeling off the purplish rind, young tall stalks are useful in various ways. They are pleasantly acid raw, and like rhubarb, are good in various recipes. Gibbons has made tasty pies, sauces and jams out of them. Roots can be boiled or baked for food.

Polygonum

This plant can be used in salads or for

persicaria
Spotted ladysthumb
(16)

flavoring purposes.

Polygonum spp.
Knotweeds, smartweeds

Seeds of several *Polygonums* can be parched and pulverized for food purposes. Throughout the world, roots of various species of bistort have been used in emergencies.

Some knotweeds have been suspected of causing photosensitivity in animals.
(12, 16, 19)

Polyozellus multiplex
(32)

This mushroom is rated as choice eating in reports that are sketchy. It should be treated with care as with any unfamiliar mushroom.

Polypilus frondosus
Hen of the woods
P. umbellatus
(32)

Young stages of both of these mushrooms make better eating than when they are older. If *P. frondosus* is tough, discard the stems. Slow cooking is recommended with this species.

Polyporus sulphureus
Sulfur mushroom
(15, 24)

This mushroom is described as excellent, but may be inclined to be tough. It should be thinly sliced, and stewed or sautéed for a while.

Populus spp.
Aspens, poplars, cottonwoods
(13, 25)

The sap or bark can be used for food. On leaves or buds, aphids often produce an edible honeydew. This can be scraped off, or released by boiling. The inner bark of at least the black cottonwood (*P. trichocarpa*) contains vitamin C; and Indians used its catkins raw for food.

Portulaca oleracea
Purslane, pusley, common portulaca
P. retusa
P. sativa
Golden purslane

The first and last entries are popular food plants, and have been cultivated for food for centuries. The foliage is popular in salads, and is excellent when cooked. It has been pickled, used to thicken soups, or dried or frozen for storage. Seeds are used in the form of flour or meal.

No cases of oxalate poisoning from *Portulaca* have been reported in North America, as has been the case in Australia. It would be worth testing for oxalate if doubt exists. (2, 8, 15, 16, 19, 24, 25)

Potamogeton spp.
Pondweeds, fishweeds
(16, 25)

Roots of some species are eaten raw. The root parts in any case can be cooked for food.

Potentilla anserina (Argentina anserina) (A. argentea)
Silverweed, cinquefoil, silverweed cinquefoil, goosegrass, fivefingers
(12, 16, 24)

North American Indians made frequent use of the roots of this plant. They are good raw, and excellent cooked.

Potentilla fruticosa
Shrubby cinquefoil, bush cinquefoil, yellow rose, fivefingers
(16)

Tea has been made from the leaves.

Primula veris
Cowslip

Foliage and flowers were much used in medieval times in salads.

(8)

Prosopis juliflora (P. glandulosa) (P. chilensis) Mesquite, honey mesquite, honeypod, kiawe bean

Green pods, cooked, are considered a delicacy by Indians, and the pulp surrounding the seeds is sweet. Mature pods can be dried and pulverized to make a flour described as highly nutritious. Some have eaten them raw. Flowers can be stripped off their central stalk and eaten as is. Secretions of hardened sap, when they can be found on the bark, are high in sugar content, as are the pods and flowers.

Cattle feeding on large amounts of this plant can undergo various conditions of poisoning, some of them irreversible (such as an inability to digest cellulose). The mechanism of poisoning is not completely understood, but might be in part a nutritional problem.
(16, 19, 24)

Prosopis pubescens Screw bean
P. velutina Arizona mesquite
(16, 24)

The pods of these plants are useful in much the same ways as with *P. juliflora.*

Prunus spp.

Some of our best domesticated fruits come from this genus, such as apricots, peaches, cherries, and so on. Wild species range from chokecherries to wild plums, and they can be utilized raw, dried, ground up, or in recipes. The seed inside the stone is worth eating, where there is one.

It should be no deterrent to moderate use of species of *Prunus*, but some of them are rich in cyanide and have proved fatal to children and to livestock. Seeds and leaves are

particularly dangerous. Cyanide is volatile and is lost from leaves when they become dry— but this could not be expected of the seeds. One should remember that while cyanides are extremely poisonous, they are not cumulative and the body will metabolize away small quantities of them. As long as the use of cyanogenetic seeds or other parts stays moderate, there is little to worry about.
(1, 2, 12, 13, 15, 16, 19, 24, 33)

Prunus virginiana
(P. melanocarpa)
(P. demissa)
Chokecherry

The bark has been used in making a tealike beverage.

The above precautions with regard to cyanide should be observed.
(16, 19)

Psathyrella disseminata
(35)

Cook unfamiliar mushrooms thoroughly.

Pseudotsuga taxifolia
Douglas fir,
Douglas spruce
(24)

Leaves have been used for making tea.

Psilocybe foenisecii
(35)

To this genus belong some of the hallucinogenic mushrooms.

Psoralea castoria
Beaverbread,
scurfpea
P. esculenta
(Pediomelum

The roots of these were an important Indian food, and were prepared in the usual ways, such as boiling, baking, grinding into meal, etc. They are edible raw and keep indefinitely when dried.

esculentum)
Common breadroot, Indian breadroot, prarie apple, prarie potato, scurfpea

One analysis of the roots of *P. esculenta* turned up the following percentages of nutrients: Starch, 70%; sugar, 5%; nitrogenous matter, 9%. Roots of *P. cuspidata* have probably been eaten, but are reported to taste bitter.

P. hypogaea
Small Indian breadroot
P. mephitica
Skunktop scurfpea
P. orbicularis
Round-leaved psoralea
P. tenuiflora
scurf pea, Indian turnip

P. tenuiflora has been reported as toxic to livestock, but foliage is undoubtedly involved in this rather than the roots. (13, 16, 19, 24)

Pteridium aquilinum (Pteris aquilina)
Bracken fern, brake, eagle fern

Young uncoiled fronds are used in many parts of the world for food. In Japan, they are boiled and dried out for storing. They can be used in soups, cooked like asparagus, or eaten raw in combination with other foods. Starch has been extracted from the root parts, and the Indians dried and ground the roots into a meal.

Livestock losses have occurred through prolonged consumption of bracken fern. The toxin involved is a thiaminase which reduces thiamine levels in the body. Deficiency of this vitamin appears to be the chief cause for the various symptoms that arise. In nonruminants, thiamine therapy brings about dramatic recoveries. Bracken hay can be rendered nontoxic by boiling it for a half hour or more; cooking the fronds for human use should bring about the same detoxification.
(2, 13, 16, 19, 24, 25)

Pyrus spp.
(24)

See *Malus spp.*

Quercus spp.
Oaks

Some acorns can be eaten raw; others are too bitter for this and must first be processed to remove the astringent tannin. Since it is soluble in water, the shelled acorns are usually crushed and leached out for a prolonged period of time with water. (Indians immersed their crushed meal in fine meshed containers into streams for up to several days.) Parboiling them can also be used to remove the bitter taste. Acorns, or possibly their shells, have been roasted and pulverized for use as a coffee substitute. An edible oil can be seperated from the acorns by boiling them; it floats to the top. One analysis of acorn meal yielded the following figures: Carbohydrates, 60%; fat, 20-25%; protein, 4.5-5.5%.

All oaks should be considered to be potentially toxic. Cattle losses have occurred through eating the leaves and other parts of the plant. Tannin will cause poisoning in animals, but in one experimental investigation, poisoning in rabbits was prevented by addition to their diet of calcium hydroxide (one part to six parts of tannin). Supplemental feedings to cattle of alfalfa have also prevented poisoning. Tannin is no doubt safe for man in minor amounts, but should be considered potentially toxic. When used in the treatment of burns, for example, the absorption of large amounts of it may result in kidney damage.
(15, 16, 19, 24, 25)

Ranunculus aquatilis
Water buttercup,
hairleaf water crowfoot
(25)

The whole plant can be cooked and eaten.

Ranunculus scleratus (R. eremogenes)
Blister buttercup, cursed crowfoot

Humans in need have boiled or parboiled the young foliage and roots for food.

Ranunculus as a group has a reputation for being poisonous, although some of its members are nontoxic or nearly so. The only toxic principle thus far recognized in *Ranunculus* is protoanemonin, and this is formed by enzymatic splitting of a glycoside in the plants, ranunculin. Protoanemonin is a vesicant oil; it is capable of producing skin ulceration, or irritant lesions throughout the whole digestive tract if eaten. Cooking or drying seems to drive it off in the case of *R. scleratus*, and protoanemonin or its dimer can be obtained by steam distillation of *Ranunculus* tissue.
(16, 19)

Ranunculus spp.
Buttercups, crowfoot species

Young flowers of buttercups can be preserved in vinegar as a type of pickle. Indians boiled roots for food, and they parched the seeds and made meal out of them.

Even though *R. scleratus* is one of the more toxic buttercups, it has still been used for food. Following the methods of detoxification outlined above, possibly any buttercup can be rendered harmless and thus fit for food.
(19, 33)

Ratibida columnifera
Coneflower, prarie coneflower
(12)

Indians used the leaves and flower heads in making a tealike drink.

Reynosia latifolia
Darling plum, red ironwood

The fruit is sweet and edible, and has occasionally been sold in markets.

(24)

Rhamnus caroliniana Indian cherry, Carolina buckthorn **R. crocea** Redberry (24)	Fruits of both of these are edible. In the case of *R. caroliniana*, the fruit is sweet and edible only when it blackens to ripeness.
Rhexia virginica Meadow beauty, deer grass (24)	Leaves are pleasing additions to salads. They taste somewhat tart and sweet.
Rhododendron nudiflorum (Azalea nudiflora) Azalea, pinxter flower (24)	An irregular growth occasionally appears on the leaves and twigs. It is a valued food eaten as is, and can be used in salads or pickles.
Rhodophyllus abortivus (32)	The carpophoroids should be present to insure proper identification of this mushroom. If free from bacterial decomposition, they are edible along with the normal fruiting bodies.
Rhus glabra (R. cismontana) (Schmaltzia glabra) Smooth sumac **R. spp.** Sumacs	The berries of any red-fruited sumac can be used for food. They are commonly made into a lemonade which many people enjoy. Squawbush berries are much used by Indians in different ways. Green or ripe, they can be eaten raw; or ground up seeds and all, and dried

R. trilobata
(R. oxycanthoides)
(Schmaltzia trilobata)
Squawbush, skunkbush, skunkbush sumac
(1, 2, 12, 16, 24, 33)

into cakes. They keep well when dry. Some Indian tribes are said to use the raw shoots of smooth sumac for food.

Ribes aureum
(R. longiflorum)
(Chrysobotrya aurea)
Golden gooseberry, wax currant, buffalo currant, squaw currant, clove bush
R. cereum
(R. inebrians)
(R. pumilum)
Wax currant, squaw currant
R. spp.
(Grossularia spp.)
(Limnobotrya spp.)
Currants, gooseberries

Any gooseberry can be used in the same ways for food that other fruits are, although some of them are rather tart when eaten raw. The flowers of *R. aureum* and *R. cereum* are laden with nectar, and make good eating raw. Indians occasionally cooked young leaves of *R. cereum* with other foods.

Eating too many of the fruits of *R. cereum* may cause sickness.
(1, 2, 12, 16, 24)

Robinia pseudo-acacia
Black locust, locust tree

The seeds can be cooked for food like beans. Young tender pods are said to have been eaten by Indians, and flowers can be dipped in batter and fried to make fritters.

Most parts of this tree appear to contain a phytotoxin

which has caused some loss of life. It is not, however, potently deadly as are other phytotoxins such as ricin or abrin. Recovery appears to be complete where nonfatal poisoning in humans has occurred, but it might take a long time. All phytotoxins are destroyed by heat, so safety is assured by sufficient cooking of the plant for food.
(15, 19, 24)

Rorippa armoracia
(Radicula armoracia)
(Armoracia armoracia)
(Cochlearia armoracia)
Horseradish
R. islandica
R. nasturtium-aquaticum
(R. nasturtium)
(Radicula nasturtium-aquaticum)
(Nasturtium officinale)
Watercress, pepperleaf
R. palustris
Marsh cress, yellow water cress
(1, 12, 15, 16, 24, 33)

The greens of all these are edible, and are nice additions in salads. Younger leaves in the case of horseradish are used, and grated horseradish roots are well known for their flavoring qualities. Watercress leaves can also be dried and ground up to make a pepperlike seasoning.

Rosa spp.
Roses
(2, 8, 12, 16, 24)

Roses are best known for food in terms of their rose hips. These fruits are richer in vitamin C than even citrus fruits are, and the whole fruit is edible. There are small hairs around the seeds which detract somewhat from eating them raw. They can be eaten green if necessary, and can be utilized in different ways for food, such as making flour out of them. Young shoots can be used as a potherb. Roots, leaves and even rose petals have been used in tea making; and rose petals are also good raw or candied.

Rozites caperata
(32)

Stalks of this mushroom are tough, and it has been recommended that they be discarded.

Rubus spp.
Raspberries, blackberries, thimbleberries

Raw raspberries and related fruits make fine eating. Leaves have been used in making tea, and shoots and stems are worth trying raw. Shoots of the red raspberry and thimbleberry have been used as potherbs.

After wilting, *Rubus* leaves should be regarded as toxic. (1, 2, 8, 12, 15, 16, 24)

Rumex spp.
Docks, sorrels

Leaves, preferably younger ones, are useful as potherbs or in salads. Dock seeds were ground by Indians into meal. Olsen recommends leaching them, presumably to reduce or remove the content of soluble oxalates. Leaves of *R. acetosella* (sheepsorrel) can be used to make a refreshing drink.

The pleasant sour taste of docks comes from soluble, toxic oxalates in the plant tissue. See under *Oxalis violacea.* (2, 8, 12, 16, 19, 24, 25)

Russula alutacea
R. delica
R. mariae
R. purpurina
R. virescens
Green russula
(24, 32, 35)

R. virescens is described as an excellent mushroom.

Ruta graveolens Rue (8)	Small amounts of rue can be used to flavor salads.
Sabal palmetto Cabbage tree, cabbage palm, palmetto (24)	Indians used the fruits for food. The large, cabbagelike bud on the top, raw or cooked, is considered delicious– but removing it kills the tree.
Sagittaria spp. Arrowheads (1, 2, 12, 15, 16, 24, 25)	Any arrowhead with tubers of sufficient size are a worthwhile source of food. Roasted or boiled, the tubers are delicious. After boiling them, Indians also set them out to dry for storage purposes.
Salicornia spp. Glassworts, samphires, pickleweeds (2, 24, 25)	*Salicornias* are often used in making pickles, or they can be cooked for food.
Salix spp. Willows (2, 13)	Young leaves or the peeled shoots are edible raw. The inner bark, dried and pulverized into meal, can be used in emergencies.
Salsola collina **S. kali** **(S. pestifer)** Russian thistle, tumbleweed, saltwort	Shoots of Russian thistle are a delicious potherb. The seeds are suggested for food, and contain, according to one analysis, 3.9 calories per gram. *S. collina* should be worth trying in the same ways as with Russian thistle.

Russian thistle is a plant to be cautious with. If there is selenium in the soil, it will absorb it. Dangerous amounts of soluble oxalates have been found in the genus in Australia.

Of less concern is the presence of nitrate. These toxins are possibly less present in the form of the plant used for food, but it might pay not to take a chance. If at all possible, an area should be tested first for selenium before using Russian thistle for food. It can be detected fairly easily. Selenium in the sample is converted first to selenious acid, by treating it with aqua regia, fuming nitric acid or a sulfuric and nitric acid mixture. Alkali chlorate or perchloric acid can be added to facilitate complete solution of carbonaceous matter, but care is required in the use of these. Since nitric acid would interfere with the latter part of the test, it must be removed. It can be destroyed by adding hydroxylamine hydrochloride or sulfamic acid, or it can be boiled off. Boiling, however, can entail the loss of considerable amounts of selenium through vaporization (which can be reduced by the addition of alkali chloride).

The resulting solution, filtrate or distillate is thereupon treated with hydrochloric acid to a minimum of 80% by volume. Red selenium is precipitated from this solution by any of these reagents:

1. Ferrous sulfate (it occurs slowly in dilute solution and in the presence of much sulfuric acid).
2. Hydriodic acid.
3. Hydroxylamine hydrochloride (after long boiling).
4. Sulfur dioxide (after long boiling).
5. Stannous chloride.
6. Zinc (it becomes coated with selenium and looks like it is covered with copper).
7. *Thiourea (nitrate or larger amounts of copper interfere with this test).

(16, 19, 24, 29)

Salvia spp.
Sages

The seeds of any sage should be tried if necessary for food. They can be parched and ground into meal. The seeds of chia (*S. columbariae*) are useful in making a popular beverage, and leaves of some sages are used for flavoring purposes.

* Recommended test.

At least one *Salvia* (*S. reflexa*, or *S. lacinata*) contains dangerously high levels of nitrates.
(8, 19, 24)

Sambucus spp.
Elderberries, elders

If the berries on an elder are bluish or black in color, they can be eaten raw or cooked. Indians dried them for storage, and they are good in pies, preserves or drinks. Elders bearing red berries are a different matter, for the berries are bitter and unpleasant. Even so, they were used for food by many Indian tribes. Flower clusters with the large stems removed can be dipped into batter and fried. The flowers have been used to make tea, and can no doubt be prepared in many other ways. Young shoots are said to be useful in many other ways.

Some people have experienced difficulties from eating the red berries. Elderberries have been suspected of causing poisonings, but other than confirmation of the presence of nitrates in *S. pubens*, there has been little or no documentation. It would no doubt be best to stick to the plant parts mentioned above.
(1, 2, 8, 12, 15, 16, 19, 24, 25)

Sanguisorba minor
Salad burnet, garden burnet, poterium
(24)

Leaves can be used in salads. They taste like cucumbers.

Sarcobatus vermiculatus
Greasewood

While still tender, twigs of this shrub have been boiled for food.

Greasewood may contain 10-22% of oxalates, and has been responsible for heavy losses in sheep. See under *Oxalis*

violacea. (19, 24)	
Sarcoscypha coccinea Scarlet cup (32)	It is not a recommended mushroom, but is referred to as pleasant eating.
Sassafras variifolium (S. albidum) (S. sassafras) Sassafras, ague tree (2, 15, 24)	Sassafras roots are famous for the tea that can be made from them. For this they are cleaned and used fresh, or dried for later use. Leaves are pleasant to eat as is, or they can be dried for use as a thickener and flavoring in soups. The green buds also make a nice nibble in wintertime.
Satureja hortensis Summer savory **S. montana** Winter savory (8, 24)	Leaves can be dried and powdered for use in flavorings.
Saxifraga micranthidifolia (Micranthes micranthidifolia) Lettuce saxifrage, mountain lettuce (24)	The leaves are prized for food in some of the mountainous parts of Pennsylvania. They are used in salads, or are very slightly cooked.
Scirpus spp. Bulrushes (12, 16, 24, 25)	Rootstalks can be peeled and eaten raw, or are good cooked. They can be dried and made into flour, and the Indians used them also for making syrup (by boiling crushed roots for several hours).

Young shoots are good eating as is, and the base of stems of the mature plant is also edible raw. Pollen can be collected for addition to other foods, and seeds can be made into meal useful in similar ways. One partial analysis on the roots of *S. validus* (softstem bulrush) turned up these figures: Sugar, 8%; starch, 5.5%; protein, 1%.

Sclerocactus whipplei
Devil's claw
(3)

The whole plant is edible roasted.

Scleroderma aurantium

Some people rate this mushroom as edible. It is not recommended, but is surely worth trying if there is nothing else to eat.

This mushroom has been responsible for nonfatal poisonings, so great care should be exercised if one has to use it for food.
(19, 32)

Sedum spp.
Stonecrops, orpines, queen's crowns, king's crowns
(2, 12, 16)

Shoots and young leaves are good in salads or as potherbs. Boiled roots can be eaten.

Shepherdia argentea (Lepargyraea argentea)
Silver buffaloberry,

Berries are eaten raw, or used in preserves or cooked in various ways. Indians ground up the berries along with the seeds and made patties out of them. They also used the berries of *S. canadensis*, which are much less attractive as

rabbitberry, silverleaf	food.
S. canadensis Canadian buffaloberry bitter buffaloberry, russet buffaloberry, soapberry	

Berries of *S. canadensis*, according to one analysis, contain 0.74% saponins, which are responsible for their bitter taste and the fact that they froth up when whipped. It would be best to go easy on the berries or avoid them for this reason.
(16, 24, 38)

Sidalcea malvaeflora Harsh sidalcea (13)	Greens of the plant can be eaten.
Silphium laciniatum Compass plant, pilotweed, resinweed (24)	The plant sap hardens into lumps used by Indian children for chewing gum.
Simmondsia chinensis Goat nut, jojoba (33)	The seeds and nuts have been roasted and ground to make a substitute for coffee.
Sisymbrium spp. Tumblemustards, hedge mustards	Young leaves and shoots can be used in salads, and are commonly used as potherbs. Seeds have been gathered by Indians and made into meal for use in the usual ways.

Large amounts of foliage have been reported toxic to livestock, but are probably nothing to worry about in normal amounts in the human diet.
(16, 24, 25)

Sium suave
(S. cicutaefolium)
Water-parsnip,
water-parsley

Roots have been reported to be edible in late autumn.

There is lack of clearcut evidence that this plant is toxic, but the upper part of it has been reported as having been lethal to cattle. Correct identification of this plant is imperative, for it closely resembles the deadly water hemlock (*Cicuta douglasii*).
(12, 19)

Smilacina spp.
(Vagnera spp.)
False Solomon's seals,
Solomon plumes

Berries are edible either raw or cooked, but the taste is nothing to get excited about. The roots are bad tasting as is, and to prepare them for food, Indians would first soak them in lye. They thereupon parboiled the roots to get rid of the lye. The rootstalks are otherwise starchy, and have been used in making pickles described as pleasant. Tender shoots can be used when cooked.

Berries may have a cathartic effect if eaten in quantity.
(12, 16, 24)

Smilax bona-nox
Bristly greenbrier
S. herbacea
(Nemexia herbacea)
(N. lasioneuron)
Greenbrier,
carrion flower,
Jacob's ladder

Tender shoots of Jacob's ladder can be prepared and eaten like asparagus, and berries of the plant are pleasant eating. Roots of the first and last entries were ground into flour by the Indians and used in various ways.

S. pseudo-china
China brier,
long-stalked greenbrier
(16, 24)

Smyrnium olusatrum
Alexanders,
black lovage
(8)

The leaves can be used in flavoring salads.

Solanum elaeagnifolium
White horsenettle,
silverleaf nightshade,
tropillo

Indians used the crumbled up berries to curdle milk into cottage cheese which was used immediately. Two berries suffice for a quart of milk.

Solanums have a reputation for being poisonous plants, and this is caused by a class of toxins called glycoalkaloids. When consumed, these cause varying degrees of irritation in the gastrointestinal tract. Further, they readily undergo hydrolysis to form sugars and alkamines, which is another class of poisons. Alkamines cause nervous symptoms in affected animals, and are what is responsible for cases that prove fatal. Poisoning from *Solanum* consists of varying combinations of the effects of these two toxins. Where death does not occur, there appears to be complete recovery. It should be noted that *Solanum* cannot be detoxified by drying. Although there are exceptions (such as the potato and tomato), it is safest to consider all species of *Solanum* as dangerous unless otherwise determined.
(16, 19)

***Solanum fendleri**
Wild potato
***S. jamesii**

Tubers of *S. jamesii*, at least, contain a bitterish taste concentrated mostly in or near the peeling. To counteract this or

* See top entry of that genus.

Wild potato
(16, 24)

other undesirable tastes in food, Indians often mixed clay in with it. These tubers can be eaten raw, boiled, baked or roasted; and they can be stored fresh like potatoes or after being dried.

***Solanum nigrum**
(S. americanum)
(S. interius)
Nightshade,
black nightshade,
garden nightshade,
deadly nightshade,
poisonberry
***S. nodiflorum**
Small-flowered nightshade

Ripe berries have been eaten raw, and are much used by some people in pies, preserves and puddings. They have a musky flavor to them, and should be used only when black, indicating they are ripe. Young shoots are used in parts of the world for a potherb.

Even the edible parts mentioned have been reported to have caused occasional sickness. One should go slow to make sure he is not sensitive to these foods.
(4, 12, 16, 19, 33)

Solidago confinis
Southern
California
goldenrod
S. missouriensis
Missouri
goldenrod
S. odora
Sweet goldenrod,
anise-scented goldenrod,
blue mountain tea

Leaves of the first two entries are recorded as being edible. While not recorded as being directly edible, the fragrant leaves of *S. odora* have been widely used in making a pleasant tea. Dried flowers of the same species are also said to be good for tea making.

Goldenrods should be regarded as potentially toxic. See under *Solidago spp.*
(13, 15, 19, 24, 32)

* See top entry of that genus.

Solidago spp.
Goldenrods

Seeds can be ground up for use in various ways.

Nitrates have been found in goldenrods, but a greater danger is posed where the plants have been infested with rusts. These fungus infestations are extremely toxic, and plant parts gathered for food should be carefully checked for such growths. Some goldenrods are otherwise suspected of being toxic; the group should be carefully approached for food use.
(19, 25)

Sonchus spp.
Sowthistles
(2, 12, 16, 24)

Young green parts have served in salads and as potherbs since at least the time of the Roman Empire. Any bitterness may be compensated for by blanching the leaves, mixing them in with other greens, or parboiling them. Peeled young stems and the roots are recorded as being edible.

Sorbus americana
S. aucuparia
Mountain ash, rowan tree
S. scopulina
Mountain ash
(12, 24)

The berries are used in preserves, and in the case of *S. scopulina* at least, were dried and ground into meal.

Sparassis radicata
(32)

This is a mushroom prized by connoisseurs.

Sparganium spp.
Burreeds
(12, 25)

The bulbous stem bases and tubers can be eaten raw, boiled, or dried and made into meal.

Spiraea spp.
Spireas
(12)

The leaves, stems and flowers of some species were used by Indians for tea making.

Spirodela spp.
Great duckweeds
(16)

These aquatics can be eaten raw if necessary.

Stachys palustris
Betony
(25)

Parched seeds can be consumed or made into meal.

Stellaria jamesiana
(Alsine jamesiana)
(A. curtisii)
Tuber starwort, chickweed
(16)

Raw, the tubers are pleasant eating, and they can be cooked in practically any way to make good food. For future use, the tubers can be dried in the sun.

Stellaria media
(Alsine media)
Chickweed, common chickweed

Chickweed shoots have somewhat of a bland taste, but are good in salads or cooked like spinach. Leaves can be dried for use in tea making.

Nitrates have been measured in *S. media.*
(2, 16, 19, 24)

Sticta glomulifera
(23)

This lichen can be prepared for food by boiling it until it coagulates like scrambled eggs.

Streptopus amplexifolius
Twisted-stalk,

Indians ate the berries, and young green parts have been used in salads.

white mandarin,
liverberry
(2, 12)

Strobilomyces floccopus Old man of the woods **S. strobilaceus** Pine cone mushroom (24, 32)	The pine cone mushroom is excellent eating, but the top entry is poor in taste.
Strombocarpa pubescens (Prosopis pubescens) Tornillo, screw bean (24)	The pods are high in sugar content. They can be chewed on, and have been used in making drinks and syrup (by boiling the crushed pods). They can also be ground into food meal.
Stropharia ambigua (32)	This mushroom is said to taste poorly, like old leaves.
Suillus americanus **S. brevipes** **S. granulatus** **S. grevillei** **S. lakei** **S. luteus** Slippery jack **S. subaureus** **S. subluteus** **S. tomentosus** **S. umbonatus** (32)	Some of these mushrooms are choice. Slippery jack and *S. brevipes* are thus described, and Smith says in referring to them that the slime and tubes should be removed from the caps before cooking. *S. umbonatus* is described as only probably edible.

Symplocarpus foetidus (Spathyema foetida)
Skunk cabbage

The young leaves can be parboiled to make good eating. During cooking the offensive odor is removed.

Roots of skunk cabbage contain irritating crystals of calcium oxalate, as in the case of jack-in-the-pulpit.
(19, 24)

Tanacetum vulgare (Chrysanthemum vulgare)
Tansy

Tansy leaves can be used in culinary flavorings.

Excessive use of medicine prepared from this plant has caused fatalities.
(8, 19)

Taraxacum officinale
Dandelion, blowball
T. spp.
Dandelions
(2, 8, 12, 15, 16, 24)

There are probably more ways to fix the dandelion up for food than there are ways that it cannot be done. Starting with the leaves, young ones can be eaten raw if no person has been out spraying things. If somewhat older, they will begin to taste bitter, which can be coped with in the usual manners, such as selecting blanched leaves or parboiling them. They are an excellent potherb. Dried leaves are said to be useful in making a tealike drink. Very young flowers, or better, flower buds, can be lightly cooked to make an excellent vegetable. (The well known dandelion wine is also made from the flowers). Although slightly bitter, the seeds can be eaten. Juice extracted from the roots is supposed to be very nourishing; the roots are otherwise useful in a variety of ways, such as raw in salads, parboiled, fried, roasted, or dried and pulverized for a coffeelike substitute. Crowns (bases of plants above the root) are delicious vegetables raw or cooked. It should be noted that the common dandelion is very high in vitamins A and C, and any *Taraxacum* should be tried if necessary for food.

Taxus spp.
Yews,
ground hemlocks

The fleshy part of the fruit is edible.

The foliage, bark and seeds of yews are very toxic, and fatalities are frequent in livestock that have eaten of the plants. Just the fleshy fruits should be eaten (for safety only in moderation), and one should not forget to throw the poisonous seed in each fruit away.
(19, 24)

Thelesperma longipes
T. megapotamicum
(T. gracile)
T. subnudum
Navajo teas
(16)

Foliage can be used to make a tea described as being very delicious. It can be dried for storage.

Thlaspi alpestre
(T. fendleri)
Blue pennycress
T. arvense
Field pennycress,
fanweed

Field pennycress is cultivated in Europe for the shoots, which are edible raw, or cooked like spinach. There is bitterness in the older parts which detracts from the pleasure of eating it, but the plant is high in vitamin content (C and G). The seeds and fruits are said to have been used for flavoring purposes.

Seeds of field pennycress contain allyl isothiocyanate, which is one of the mustard oils. In quantity or concentration, it will cause disturbances in the digestive tract, and 0.001% of an animal's weight in oil is lethal to cattle. There is, however, no cumulative effect; and just as with mustard seeds, no harm should come from using them moderately. (16, 19)

Thymus spp. Thymes (8, 24)	Leaves can be used in flavorings.
Tilia spp. Basswoods, lime-flowers (8, 15)	Dried flowers can be used to make a very pleasant tasting tea.
Togaria aurea (32)	Some people may suffer digestive upsets from this mushroom, but it is otherwise edible.
Tragopogon dubius Salsify **T. porrifolius** Salsify, purple oyster plant, **T. pratensis** Meadow salsify, goatsbeard, oysterplant (12, 16, 24, 38)	*Tragopogon*, as its common name would suggest, tastes somewhat like oysters, and *T. porrifolius* has been cultivated in many places for food. The roots of these species are edible raw, or can be prepared by cooking them in any of the ways that are possible with potatoes. (*T. porrifolius* is recommended over the other two). Young shoots can be eaten raw or cooked, with parboiling if the taste is too bitter. The coagulated juice was used for chewing gum by Indians. The tops of *T. porrifolius*, at least, can be used for greens.
Tricholoma equestre Canary mushroom **T. personatum**	Unfamiliar mushrooms should always be well cooked.

T. russula
T. sejunctum
T. transmutans
(35)

Trifolium spp.
Clovers

Clovers are plants high in protein content, and are useful for food in various ways. Various species were eaten raw and cooked by Indians, and produce edible seeds, roots and young blossoms. Flowers and seeds in some cases have been dried and ground into meal. The blossoms and greens of *T. pratense*, at least, have been used in making a tealike drink.

Different clovers have been found to cause different types of poisoning in animals, ranging from photosensitization and hepatic damage to poisoning by cyanide, and beyond. In all cases, commercial strains of clover under farm cultivation have been responsible for the trouble. Thus clovers growing free from the hand of man are presumably safe for food, or at least safer.
(2, 12, 13, 15, 19, 24, 25)

Triglochin maritima
Arrowgrass, podgrass, shore podgrass, seaside arrowgrass, goosegrass, sourgrass

The leaves are reported to have been used raw in salads (not recommended!), and cooked as a potherb in some localities. Indians parched the seeds for food and for a coffeelike substitute.

This plant owes its toxicity entirely to hydrocyanic acid, which is apparently not present in glycosidic form as in most plants which are cyanogenetic. This cyanide is lost from the plant upon heating or prolonged drying, but one should still be aware of the danger. One should not eat the plant raw.
(12, 16, 19, 25)

Trillium grandiflorum Large-flowered trillium, trillium, wake robin **T. ovatum** Pacific trillium, birthroot **T. spp.** Trilliums (12, 24)	These and other species of *Trillium* are recorded as being good potherbs.
Triosteum spp. Wild coffees (2, 24)	The berries of at least three species have been dried, roasted and ground for use as a coffeelike substitute.
Tropaeolum majus Nasturtium (8)	Leaves are useful in salads, and the green seeds can be used as capers.
Tsuga spp. Hemlocks (2, 24)	Tea can be made from the leaves, and raw shoots are edible.
Typhaceae Cattail family (2, 12, 15, 16, 24, 25)	Cattails must surpass all other plants in the number and variety of foods that can be prepared from them. The underwater root parts are starchy and nutritious, and are available at any time of the year (but seem to be richer in starch toward autumn). They are pleasant eating raw, except for fibers in the root cores, and are good when boiled or baked. Flour can be obtained from them by drying and pulverizing the core of the root, then sifting it to remove the fibers.

There is also a water method for producing flour, where the root cores are broken down and mechanically manipulated in cold water. The starch readily settles to the bottom, whereupon it can be washed and the fibers strained out. This water method produces a better tasting flour, according to Gibbons. Cattail shoots are available mostly in the spring, but do appear through most of the year. They are excellent eating. After peeling them, they can be used raw, or cooked in a variety of ways. Young flower stalks can also be cooked in different ways. One way is to eat them like corn on the cob. Later on, the pollen producing flowers can be mixed with flour for use in making bakery goods, or they can be preserved for future use by drying them. The pollen itself is well worth collecting for food, and can be used like the flowers in cooking. Indians also utilized the tiny seeds for food. The fluffy part was first burned off.

Ulmus fulva
Slippery elm,
red elm
(24)

The inner bark of the tree or the root are pleasant eating raw, or can be made into flour.

Umbellularia californica
California bay tree,
California laurel,
Oregon myrtle
(13, 33)

The fruits are edible when roasted.

Umbilicaria dillenii **U. muhlenbergii** **U. pustulata** **U. vellea** Rock tripes (24)	Rock tripes are first thoroughly cleaned, then slowly roasted until crisp. They should next be boiled for about one hour. Rock tripe is described as tasting somewhat like tapioca with a hint of licorice.
Urtica dioica **(U. gracilis)** **(U. viridis)** **(U. procera)** Stinging nettle, slim stinging nettle **U. spp.** Nettles (2, 8, 16, 24, 25)	Young greens are much used as a potherb in parts of Europe. They are bland in taste; cooking completely removes the stinging qualities of the plant. When picking them, include the pink stem down to the root. Nettle leaves are also used in making tea; the plant is supposed to be high in vitamin C. Any nettle should be worth trying for food.
Uvularia perfoliata Perfoliate bellwort, wild oat (24)	The roots and plant shoots can be eaten when cooked; the shoots are described as excellent eating.
Vaccinium spp. Blueberries, huckleberries, whortleberries, grouseberries, bilberries (1, 2, 12, 15, 16, 24)	The berries are choice fruits raw, in preserves, in recipes, or probably any other way. Indians dried them to a raisinlike condition for storage. Leaves are useful fresh or dried for making a beverage.
Valeriana edulis **(V. obovata)** Edible valerian,	The root of edible valerian is rather offensive in taste and smell to some people, but is very nutritious. (The odor

tobacco-root
Valeriana spp.
Valerians

of the root has been described by some as being like that of unwashed feet, and it is said to taste much like chewing tobacco.) It has been cultivated for food in the eastern United States and was extensively used by the Indians. They baked the roots in fireless pit cookers for about two days. The roots have also been dried and made into meal. Other valerians have similar smelling roots and should be tried for food if necessary.

Indians have described the raw roots as being very poisonous.
(1, 12, 16, 24)

Valerianella spp.
Cornsalads
(8, 24)

The greens of some species are much used in salads and as potherbs. All members can be used in the same way.

Veronica americana
American speedwell, American brooklime
Veronica spp.
Speedwells
(16, 24)

American speedwell is described as being high in vitamin C and otherwise healthful. It is useful in salads, but is said to have a touch of bitterness. Any *Veronica* should be tried.

Verpa bohemica
Early morel

It is a nicely flavored mushroom, but should be well cooked.

This mushroom should only be eaten in moderate quantities. A few people eating much of it have experienced a lack of muscular coordination.
(32)

Viburnum

Fruits of these *Viburnums* are generally

alnifolium
Hobblebush, tanglefoot, American wayfaring tree
V. cassinoides
Appalachian tea, withe rod
V. edule
(V. pauciflorum)
Squashberry, mooseberry, viburnum, arrowhead
V. lentago
Sweet viburnum, nannyberry, wild raisin
V. nudum
Larger withe rod
V. opulus
(V. trilobum)
Cranberry tree, high-bush cranberry, pimbina
V. prunifolium
Black haw, stagbush
(2, 16, 24)

pleasant eating when they are raw, and many of them have been utilized in preserves, sauces, and in other cooking applications. The leaves of *V. cassinoides* have been dried for use in tea making.

Vicia spp.
Vetches

The seeds and young stems of vetches have been eaten by people throughout the world.

On a worldwide basis, vetches have been found to contain different toxic principles, such as cyanogenetic glycosides, beta-cyano-L-alanine, and principles causing photosensitivity and liver lesions. Several species have been found to be toxic when they grow in other parts of the world, but not when they grow in the Western Hemisphere. Two species mentioned as being dangerous in this hemisphere are *V. faba* (fava bean) and *V. villosa* (hairy vetch). All of them,

however, should be held suspect until demonstrated otherwise. *Vicias* have often served as food plants, but one should not take a chance by eating excessive amounts or eating the plants raw.
(1, 12, 19, 24)

Viola spp.
Violets
(12, 16, 24)

Violets, in particular aquatically inclined ones or blue ones, should be tried for food if need be. They can be used in salads or in cooking. With *V. americana*, at least, there is a bitterish taste to it.

Vitis spp.
Grapes
(15, 16, 24)

Wild or domesticated grapes can be used in much the same ways. Besides the grapes, the leaves can be eaten raw or cooked for food. Young tendrils are also edible raw.

Volvaria bombycina
(35)

Unfamiliar mushrooms should be well cooked.

Washingtonia filifera
California fan palm
(24)

Indians ate the fleshy part of the berries, then made the seeds into food meal.

Wisteria spp.
Wisterias
(15)

Flower clusters can be dipped into batter and fried to make fritters.

Wyethia spp.
Mulesears

Indians are said to have subjected the roots to fermentation for one or two days before cooking them. Olsen says the roots are edible after merely long cooking. Seeds can also be eaten.

The leaves, if nothing else, are reported to be poisonous. (12, 25, 33)

Xanthium strumarium
Cocklebur, clotbur

The seeds inside the burs are reported to have been eaten raw by Indians or ground into meal. They are pleasant tasting.

Hydroquinone is present in seeds and young shoots of the plant. It causes pathological changes in the liver and kidneys of affected animals, as well as internal bleeding. Nothing is said of the condition of the Indians who ate the seeds, but it would be well to be aware of the dangers involved in eating them. Perhaps hydroquinone is a non-cumulative toxin. If so, it would probably be all right to eat the seeds without ill effects in very moderate quantities. This should be found out.
(12, 19)

Yucca baccata
Yucca, datil yucca, Spanish bayonet, Spanish dagger, soapweed
Y. glauca
Yucca, small soapweed, Spanish bayonet, Adam's needle, bear grass
Y. spp.
Yuccas

Whether they are of the types that dry up or remain fleshy at maturity, yucca fruits are good sources of food. The ones that dry at maturity are good for food while still unripe. The skin of these may be somewhat bitter, so they can be peeled, then boiled until tender, or just eaten whole. Indians also baked them, or dried them for storage. The ones that are fleshy when ripe are good for food anytime. The pulpy part was used in various ways in cooking, such as baking, roasting, or boiling it down to paste. The seeds of this latter kind were roasted for eating, or roasted and ground into meal. Young emerging flower stalks of either kind can be roasted, then eaten or dried for future use. Flowers can be eaten raw or boiled. If the taste is bitter, try parboiling them.

The parts of the plant that lather up contain saponins, which are toxic. They should either be avoided for food, or crushed and washed, parboiled or otherwise cleansed of this toxic principle.
(3, 16, 24)

Zamia floridana Florida arrowroot, coontie, **Z. pumila** (24)	The roots were an important source of food for the Seminole Indians. They are starchy and can be made into flour.
Zizania aquatica (Z. palustris) Indian rice, wild rice, water oats (2, 15, 24)	This plant furnishes the wild rice which is the expensive gourmet item found in markets. It was a very important food item for Indians in the North.

Wintertime is naturally when the greatest problem will exist in getting fresh greens for table use. There is a technique which can help to alleviate this, called forcing. Such things as dandelion roots can be dug out and put into something portable, such as cans or boxes. Freezing the roots fools the plants into thinking winter has already occurred, and they start putting out, in some cases, copious amounts of greens. This can occur over a substantial period, and the greens are trimmed off for food as desired. These are some of the plants with which this has been successfully done:

Barbarea verna & *B. vulgaris*	(Winter cress)
Cichorium intybus	(Chicory)
Phytolacca americana	(Pokeweed)
Rumex, about 15 different species	(Docks)
Taraxacum officinale	(Dandelions)
Urtica dioica	(Stinging nettle)

There may well be many other plants with which this would work; any plant producing edible greens should be worth a try, especially ones with large roots.

When everything else has failed, an individual may still have access to abundant sources of food. Harrington (16) refers not only to algae and aquatic plants of rivers and ponds as being edible, but says that the mud and silt of ponds and lakes can be cooked into a type of soup. While aquatic plants may be edible and nutritious, one should take care to be aware of at least the one great exception to this, the deadly *Cicuta douglasii* (water hemlock). Cooking mud or silt for soup might not be as radical or far out as it sounds, for these are usually teeming with populations of protozoa and other small creatures. Where this was the only source of food for a segment of a starving population, steps would have to be taken to insure that the creatures were not wiped out or seriously reduced through overly rapid removal of the strata containing them. Protozoa typically encyst to survive during wintertime, and during this period, this would perhaps mean that they were even less favorable sources of food. Probably the gravest danger posed by these sources of food would be in bodies of water which have been industrially polluted. In a sense, it would be a form of justice if the people responsible for pollution had to ultimately turn to their own lakes and rivers for food.

There probably will not be any need for insecticides during a famine. Insects are good to rich sources of protein which humans should come to appreciate and look forward to.

Insects are a common staple item in Africa, and termites in particular are eaten with relish. The story is told of one hunter who during a hunt, was so impressed with the enthusiasm with which the natives were eating termites that he decided to try it himself. He popped a tender white one into his mouth, and let out a cry of pain. The termite had bitten his tongue, and would not let go. This was quite hilarious to the natives, who informed him that one should part the head from the body first.

Unquestionably, the great obstacle to eating insects or other new things is psychological, just as children make up their mind they do not like a food even though they have not tried it. Having had no contact with civilization, a native from the depths of bush country could well feel the same distaste and hesitation if we offered him a Hostess cupcake. During actual famine, this factor may not matter too much. There is no equal to hunger for whetting the appetite.

When the need arrives, one should not hesitate to gather whatever insects he can find; moths, mosquito larvae, flies, earthworms– anything. A person will be all the better off if he knows how to encourage insect growth. He can then stand by and reap as desired. Any study of insects and where to locate them, and so on, can only reap dividends. The method of food preparation will depend much on the insect. Locusts are easily fried, and eaten whole after the head and internal organs have been pulled off (or just eaten whole). Other meaty species should be likewise easy to prepare.

For insects where there is no great quantity of protein, where there seems to be mostly superstructure or body fluid, the whole mass should be put under pressure and the runoff filtered and gathered. This will contain the desired proteins. (Nothing here should be hard to improvise. A piece of cloth can make a suitable filter. If he had to, one could roll a bottle over the insects to apply pressure. The main problem will be one of gathering enough fresh insects to make the method worthwhile.) Thereafter, one proceeds to seperate the proteins from impurities in the runoff. Heating will coagulate most proteins. The coagulant is then cleaned and readied for the table. Proteins can also be precipitated by drying, or by adding some concentrated electrolyte to the runoff, such as a nontoxic salt, base or acid. Some of these methods may give better results than others. For example, heating will not precipitate all proteins, but would tend to be disinfectant in its action. An individual should have little difficulty in trying various methods and picking a favorite.

SURVIVAL AS A GROUP

Moral groups of the kind referred to at first should properly not be having to defend themselves against each other. Cooperation will be what saves lives, and this applies to groups as much as it does to individuals.

Whatever the situation, groups should be firm in what measures they take for protection. Not everyone is going to belong to an orderly, moral group. Nice horizons may lie ahead for the group that can keep from losing its equipment to mobs. Things will stand or fall on this ability to protect.

In ascending order of equipment requirements, let us proceed to list processes open to a group whereby it can make food. The further toward the end that an option is listed, the more sophisticated it is. This will mean the group must be larger to support it, and in particular, that men of intelligence and training be part of that group. Based on traditional standards, some of these processes may rate as uneconomical or costly, but this will have little meaning where human life is at stake. They will furthermore take time to set up and work the bugs out of; during the famine itself is not when to start thinking about them.

In the area of land under group protection, a group should pay careful attention to what plants can give in quantities of food and nutritional value, and how fast they can give it. For a particular area, highest crop yields may be obtained from the potato. In others, the potato may be inferior to other food plants. Along with everything else to take into account, at least two native plants are worth considering as excellent food sources:

1. Cattails (*Typha*). Considering just the rhizomes, it has been reported by Classen (10) that an acre moderately thick with cattails will yield around 5,500 pounds of quality flour. Classen speaks of it as being comparable to wheat flour in quality, and one should realize that in addition to this, most other parts of the cattail are edible and pleasant. Data should

be gathered on other pertinent aspects of cattails, for they have truly great crop potential. It would be an excellent way of utilizing otherwise worthless swamp areas. In fact, where other crop production was not up to the par of cattails, an area could be inundated and cattail growth begun.

2. Dandelions (*Taraxacum*). These have a long history of benefiting man nutritionally. Perhaps no one in the group is a farmer, but not much skill would seem to be involved in raising dandelions. Most parts of the plant are edible, and picked at the right time and properly prepared, it is choice and delicious.

Group control may be assumed over a natural area, where it is decided to harvest just what grows wild, rather than making the whole thing into a domestic farmland. Where this is the case, some farm practices could still pay off nicely. Enough of whatever is picked should be left behind to ensure a sustained supply, and the best specimens should be left for this. It may even pay to attempt breeding of better strains, even though the plants are wild: Normal crop plants are no match for many of them in their native environments.

In a lengthy famine, a group with its control confined to city limits should have no qualms about removing the asphalt covering on roads, if that would significantly increase the area they can grow food on. Such soil might take a little while to return to normal, but it might be excellent for direct use in growing aquatic crops. Where food crops have been carefully chosen, a group will be cultivating familiar crops in some cases; in others, it will be cultivating cattails, dandelions, or other high yield "weeds." No group should make the mistake of deciding that just because something is a traditional food item, it is going to raise some crop that is sensitive, or poor in yield.

The realities of seperate groups and individuals having to live with each other in a region should play a heavy role in the areas which they choose to control. Free areas and corridors should be established for individuals and other groups. In this way the situation could be averted, where one could not enter perhaps a mountainous area in search of food, because some narrow-minded group would not let him through an ill-conceived piece of land blocking his way.

The value of a food protein such as milk or meat lies in the fact that it is broken down in the stomach into amino acids. Proteins would be of no importance nutritionally if they did not yield amino acids; these are what the body needs and can use.

These amino acids happen to be obtainable from any protein, whether it is a food or a structural protein. Lacking enough food proteins in a famine, we can supplement our sources by turning to proteins that man does not usually consider as food. Included here are such things as hair, fingernails, toenails, whiskers, hooves, skin, leather, feathers and horns. These are commonly referred to as keratins, and by definition, have the amino acids needed by the body for nourishment. Silk is another fibrous protein which can be used. Obviously there are going to be limits to the amount of hair, fingernails or feathers which people can gather; but if after 120 days, a group has collected 35 pounds of such proteins, then it is that much further ahead of the group that has not. It should at least be comforting to know that our wool coats, hunting trophies and leather shoes offer some protection against starvation.

The hungry hoboes in the depression days of the 30's who were trying to boil boot leather for food were on the right track, even though just boiling it probably did them no good. Through hydrolysis similar to that performed in one's stomach, all proteins can be converted into amino acids. One takes something like hair or feathers, performs hydrolysis, and prepares the resulting amino acids for consumption. In the case of any protein, this will yield forth a sizable number of amino acids needed by the body. Some amino acids are possibly missing from some of these fibrous proteins, but even improper nourishment is better than none at all.

Taking wool as an example of a keratin, it is made up of 20 different amino acids (Sweetman, 34). This is about as many as the average food protein contains. The only essential amino acid apparently not present in wool is tryptophan, but seven out of eight of the essential amino acids is not bad for a start.

Hydrolysis can be performed by treating a protein with acid, base or enzyme. Enzymatic hydrolysis gives amino acids free from degradation, but it does not proceed to completion; and is not used for this reason on a commercial basis. Keratins are easy to convert into amino acids using acid or base, more so at least than with other proteins. Whereas food proteins are boiled by one method for from 16 to 24 hours, keratins require about eight. Acid hydrolysis is the least destructive to amino acids, and hydrochloric acid of around 20% strength is most commonly used for this. The time required for hydrolysis may be greatly shortened by any of these methods:

1. Raising the temperature (not to exceed 180° C).
2. " " concentration of acid or base.
3. " " ratio of acid to protein.
4. " " pressure (not to exceed 80 p.s.i.).

Since the percentage of recovery of hydrochloric acid is not dependent on concentration, there would seem to be no advantage in not using the highest convenient concentration. Further, this would permit lower temperature hydrolysis. (This yields amino acids superior in flavor to those produced at higher temperatures.) It should be pointed out as well that if hydrolysis is terminated before its completion, intermediate peptides will also be present. Perhaps peptides are adequate for nutrition as is. If so, this would further shorten the process.

Hydrochloric acid is better than sulfuric acid for hydrolysis, because of its volatility and ease of recovery from the amino acid mixture. The bulk of it can be recovered by vacuum distillation. High vacuums (to 0.1 micron of mercury or less) would give better recovery, but such units can cost from around $150 up, at 1970 prices. A simple aspirator can still give a fair recovery, and is available for as low as $2. For what of the hydrochloric acid cannot be distilled and is combined, it is simply neutralized with sodium hydroxide to give table salt, which is a dietary need anyway. If sulfuric acid is all that is available, it can be removed from the reaction mixture by neutralizing with calcium or barium hydroxide, if either is available. It possibly could be put to better use by reacting it with table salt to give the more versatile hydrochloric acid.

One problem is encountered in acid hydrolysis. Tryptophan is completely destroyed by it under normal conditions; and cystine and tyrosine undergo partial decomposition. This, along with the "humins" which are formed, can be minimized or avoided in the absence of oxygen or in the presence of a reducing agent, such as urea or stannous chloride. Where tryptophan is unavoidably lost in this way, it can be replaced by using the process of basic hydrolysis. Except for tryptophan, this method results in the racemization and partial destruction of amino acids, which is why it is not used more in protein hydrolysis.

Acid hydrolyzates tend to have a meat broth, or burnt taste to them, depending on the conditions of reaction. Elevated pressures and temperatures favor this burnt quality. Most or all of it can be removed by treatment with decolorizing carbon. If the taste of the final product is at all objectionable, it can be conveniently masked with flavorings or other foods.

Plants that cover our mountains and plains, the ones that man most commonly is in contact with, are ones which contain cellulose as the main part of their plant structure. With proper equipment and a sufficient supply of such plants, famine again is something which cannot be absolute in the threat that it poses.

Cellulose from any plant can be hydrolyzed to form sugars. This is a process which has been commercially developed in some countries, and has been used to produce vast quantities of chemicals and materials for other industries. Some of the European developers were not even interested in acquiring new sources of food; all they wanted was a cheap source of alcohol. Toward the end of World War II, Germany alone had 22 plants operating on this basis to provide yeast supplements for the diet of Germans.

Commercially speaking, the most economical form of cellulose because of its density is wood, but other forms of cellulose, such as cornstalks, bagasse and other crop residues, have been used on a limited basis. Gathering the plants up for processing is potentially the most expensive factor in the cost of commercial operations, and has been a limiting factor in the use of crop residuals. Equipment requirements can be fairly modest. As in the hydrolysis of proteins, hydrochloric acid is needed, plus vacuum for its recovery. The cellulose is hydrolyzed by dissolving it at room temperature directly into the acid. This results in the formation of intermediate polysaccharides, which after removal of acid, are converted into sugars by boiling in water (for about eight hours).

For this process the hydrochloric acid is allowed to digest the cellulose for five hours. It should be at a recommended strength of 42-45%, which does however create problems. The vapor pressure of 42% HCl at 20° C is 709 mm Hg; it is 900 mm at 25° C. This is modest pressure to have to deal with, but it does necessitate something more than simple containers, where the hydrochloric acid would simply vaporize. Furthermore, the highest concentration of commercial hydrochloric acid normally available is in the neighborhood of 37%, which means that it must be concentrated up to the 42% level. Yields in the process are excellent; cellulose which is pure has been found to give up to 97% yields of sugars. Using pine wood, commercial production has given 85% yields of such sugars as D-glucose, D-mannose, D-xylose, D-galactose and D-fructose. Pentoses and isomaltose are also produced, depending on conditions. These sugars are not uniformly fermentable; much will depend on the strain of yeast selected to grow on them. Some yeasts are incapable of fermenting certain sugars.

By increasing the time of reaction or temperature, a lesser and more easily handled concentration of hydrochloric acid might work as well as the 42%. The risk of using heat to effect hydrolysis is that sugars may be prematurely formed and destroyed by the acid, but this should be graphed out.

There is another process that has been commercially developed for the hydrolysis of cellulose, the Scholler Process, It involves the use of dilute acid, such as 0.2 to 1% sulfuric acid, and high pressures and temperatures (from 35-200 p.s.i., and 170-180° C). The sugar as it is formed is continuously removed to avoid decomposition. The trouble with this process is that high performance equipment and controls are needed, which is very possibly out of the question in famine conditions.

In any case, just dead and dried up plants should be used. Going into the process, the nearer to pure cellulose that they are, the less there will be of side reactions and side products. There would be advantage (though perhaps minor) in boiling or leaching the plants beforehand in water; it gets rid of more of the plant contents which are not cellulose, but one would thereby obtain an infusion which may be an excellent nutrient medium for growth of edible microorganisms. If live plants were to be used on a large scale, it would not only endanger their life cycles, but it would also introduce prohibitive quantities of water into the system.

If it is desired to use the resultant sugars for eventual table use, purification procedures will have to be worked out to eliminate undesirable chemicals and impurities. There would be more nutritional benefit in going to the next step anyway, that of growing yeast in the sugar solution, then harvesting the yeast for table use.

During our discussion of native plants in the preceeding section, various methods were pointed out whereby poisonous plants could be detoxified and thus rendered fit for human use. Thus we saw that heating would make the seeds of black locust or of arrowgrass fit for eating. In this discussion, we are merely making an extension of this, for with man's developed state of knowledge, there is possibly no poisonous plant which cannot be similarly made fit to eat.

Are some poisonous plants worth trying to detoxify? Yes, if the need should arise. In greater or lesser degree, all plants, poisonous ones included, contain proteins, carbohydrates, fats, vitamins and minerals; everything the human needs for consumption. Naturally there is going to be uneven distribution of these; most leaves are very low in fat content, for example. These are nevertheless normal constituents of plant cells.

Horsechestnuts are an example of multiple tons of a plant product not used because they are poisonous. The principle, if not the only, toxin in horsechestnuts is a glycoside, aesculin. It should be fairly simple to work out an extraction procedure involving the use of some solvent or hydrolysis of the glycoside.

To show the potential of this approach, it would undoubtedly be easy on a large scale basis to process poison oak or poison ivy into food. The chemical principles reported as poisons in these are soluble in alcohol and other common organic solvents. It would not be out of the question to thoroughly crush the plant tissue and extract the vesicant with solvent. Nutrient yields might be inferior compared to something else, but such things can be done.

The great majority of plant poisons are organic in nature, and in fact the only widespread inorganic toxin that is liable to pose problems is selenium. (Man's parades of insecticides, herbicides and fungicides will no doubt create problems in specific localities. Copper sulfate is a common fungicide which can render plants toxic, and plants near highways tend to have a higher lead content.) Other inorganics, such as nitrates or molybdenum, are noncumulative and are therefore not a problem under normal circumstances.

Conditions will vary with each species of plant, and let us therefore outline general procedures that should be followed to effect detoxification:

1. Because of the nature of the toxins involved, the bulk of our efforts will center around organic extractions. To successfully proceed, it is necessary to have some kind of grinding machine to reduce the plant material to a finely pulverized condition. It should be fine enough so that the toxins are no longer safely imbedded in the plant structure, inaccessible to the method applied. The majority of organic compounds are soluble in at least something, and the foremost technique for removal of poisons will be solvent extractions. Any number of solvents are worth trying, such as alcohol, ether, acetone, benzene, ethyl acetate, chloroform, pyridine, dioxane, etc.
2. Some of these solvents are expensive, but economy in any case dictates that these extractions be carried out in closed systems where they are not simply lost to evaporation. Such a setup is simple. Solvent is introduced into a container holding the pulverized plant tissue. If vapor pressure of the solvent permits, heat might be applied to hasten solution.

The solvent is thereafter removed through a filtration device, by application of pressure or vacuum. Distillation procedures are thereupon applied to both fractions, resulting in essentially complete recovery of the solvent.

3. Where there is no proper solvent available for a given toxic principle, attempts may be made to alter it to another form, which may then be more easily extracted or even thus rendered nontoxic. Heating, fermentation, change in pH, chemical reaction, prolonged storage, or drying may be factors in such changes.

4. If after the above steps have been taken, a plant exists that is still not fit for use, it may still be worth approaching from the opposite direction, that of extracting the nutrients. Fats are easily extracted or seperated, and most proteins are soluble in simple solvents, such as water, alcohol or ionic solutions with the pH adjusted. Many carbohydrates are water soluble.

The need for working things out with each plant species should be evident. Complications can arise in unexpected places, such as was found with the innocuous soybean plant. Oil has been commercially extracted from soybeans with trichloroethylene, with the leftover meal used in cattle feed. Because of this method of extraction, reactions occur in the otherwise harmless meal that have given it a well established reputation for toxicity. Other difficulties may arise as with some of the vitamins. They may be extracted in some cases along with the toxins.

Even after a toxic principle is extracted from a plant, other poisonous fractions may still be present. If so, then a succeeding extraction may be necessary. Laboratory animals should be on hand for use in feeding experiments. Where such animals are not available, the old Indian method of trying it or starving to death is an alternative. The extracted plant will at least be less toxic than it was at first.

If there is extensive growth of some poisonous plant in a region or locality, procedures should be worked out in advance for its detoxification.

Yeasts, algae, molds, bacteria, and other microorganisms

have long been considered to be potential food sources. Much research has been done on raising algae for food, and supplements of yeast have long been available to the public for food use.

Microbes are probably unsurpassed in their food potential, and as one would expect to find in nature, are part of the normal food supply for other creatures, who in turn are food for something else, etc. We can see how really excellent the nutritional possibilities of microbes are by looking at the ranges of nutrients in the dry weight of three yeasts (*Saccharomyces cerevisiae, Torulopsis utilis* and *Rhodotorula gracilis*):

Nitrogen compounds	13-59%
Carbohydrates	24-37%
Lipids	2-60%

The nitrogen compounds represent such things as combined and free proteins, peptones, polypeptides, amino acids, and many of the B vitamins (Pelczar, 26).

Other findings are equally fertile, such as when we look into the constituents of bacteria on the cellular level. Considering the cell membranes of bacteria alone, they average 50% protein, 25% carbohydrates, and 25% fats (Salle, 30). The parts within the cell are a similar story. Under optimum conditions, some microbes can multiply up to in the vicinity of a hundred times a day, which gives us reason enough in itself to consider them for food. Generally, fungi and bacteria present problems in growing and harvesting. Fungal mycelia interfere with mass cultivation, and bacteria are difficult to harvest because of their small size.

For the circumstances a group will encounter under famine, the single most important determining factor of what to raise for food microbiologically will be what man is able to gather up to feed the microbes. No group should go about it by saying something like, "Well, let's raise some Roquefort mold now." If there is little or nothing to grow it on, this or some other desirable microbe is simply out of the question. For a group having an adequate supply of sugars, it could possibly do no better than to raise common brewer's yeast (*Saccharomyces cerevisiae*), or some other better tasting yeast for food. They are easy to grow and harvest, and from the microbiological angle, the group's needs would be well met.

Sugars can be chemically derived, but circumstances may of course radically vary from group to group. The chance of having enough sugar for this last option will no doubt turn out for some groups to be remote. Where this is so, a group should take stock of what else there is to culture microorganisms on, and how big the supply is.

If for a substrate, there is no microorganism recognized as being most fit for human consumption, the group or groups should carry out research to find out what is. This will involve laboratory procedures, such as the isolation of pure cultures, their identification, exploration of toxic qualities using test animals, food analyses, and so on. Obviously nothing could be more desirable than cooperation between surviving groups, to enable more complete laboratory facilities for these and related studies.

If a group has sufficient lab facilities and enough time to use them, we should not have to hear that they starved to death because there was nothing around to grow things on. Things like meat or yeast extracts are good nutrient media for many microorganisms, but may be better just as is for human nutrition. Plants, under famine conditions, will be the best general source of nutrient media. Some protozoa, for example, grow well on hay infusions. Oak leaf infusions support some types of bacteria. If a substrate is good for nothing else, molds can usually be found that will grow on it. There is no plant, poisonous or otherwise, which should not be tried if necessary. However, there is always the possibility of carryover of toxin into the structure of a microbe from a plant. Things must be individually worked out with each plant species.

Even if there were no plant life, there are things to grow microbes on. Work has been carried out on the use of petroleum as a nutrient source. Algae are easily grown in lab mediums consisting simply of potassium phosphate, magnesium sulfate and calcium nitrate. The nutritional requirements of some other microorganisms are even simpler than that. Ordinary surface waters may contain part or all of what is needed, or infusions from garden soil may meet the same requirements. An excellent medium for growth of many types of microorganisms is urine, and solid animal wastes are common in formulas for growing protozoa.

One special group of microorganisms deserves seperate study, and they are autotrophic bacteria. Morphologically, they are no different than other bacteria, which means they are as potentially rich in food value. Since they grow on the simplest of materials, they have to synthesize their own vitamins and other nutrients. As with some algae, all they need for growth are inorganics. Some autotrophs are retarded or harmed, in fact, by organic nutrients which other bacteria thrive on. Winogradsky, the great pioneer in the field of autotrophic bacteria, was consistently able to nitrify ammonia in simple solutions containing one or two other salts. Sulfur bacteria are likewise elementary in their needs. All that autotrophs categorically require for nutrition are carbon dioxide, usually supplied as a carbonate, and some energy source, such as a simple compound of nitrogen or sulfur.

Autotrophs are difficult to grow on a large scale basis; their value lies in what they can grow on. Once the problems of production have been worked out, there are nice possibilities. Limestone or other carbonates may be plentiful in specific localities. Gypsum, to name one abundant energy source, is the wall material in many types of housing construction. It would be a shame to have to tear the walls out just to grow food to eat, but at least this option exists. Autotrophs do not require sunlight as algae do.

It should also be mentioned that bacterial slime layers, which are commonly encountered in cultures, are a possible source of food. They are usually pure polysaccharides, and the problem in using them for food would consist of finding a bacterial strain which would yield polysaccharides which would be digestible to humans.

After the isolation of a desirable food strain in any of these above cases, demands on group facilities will shrink to whatever is needed to maintain sterile conditions during mass production of the microorganism for food. If a microbial strain has been established as fit for human use, it should be passed on by a group to others that are in a position to use it. Pure cultures are also obtainable from the American Type Culture Collection, 12301 Parklawn Drive, Rockville, Maryland 20852. Universities, as well, have limited microbiological collections.

BIBLIOGRAPHY

1. Andersen, B. A. & A. H. Holmgren:
Mountain Plants of Northeastern Utah, Circular 319, Utah State University Extension Services, Logan, Utah.
2. Angier, B:
Free For The Eating, Telegraph Press, Harrisburg, Pennsylvania, 1966.
3. Armer, L. A:
Cactus, Frederick A. Stokes Co., 1934.
4. Arnold, H. L:
Poisonous Plants of Hawaii, Charles E. Tuttle Co., Tokyo, Japan, 1968.
5. Atherden, L. M:
Bentley & Driver's Textbook of Pharmaceutical Chemistry, Oxford University Press, London, 1969.
6. Beijerinck, M. W:
"On *Spirillum desulfuricans* as The Cause of Sulfate Reduction," *Centralblatt für Bakteriologie* (New Section 2), I (1895), 1-9; 49-59; 104-114.
7. Bold, H. C:
"The Cultivation of Algae," *The Botanical Review*, 8(2) 69-138, Feb. 1942.
8. Brownlow, M:
Herbs and The Fragrant Garden, McGraw-Hill, New York, 1957.
9. Burton, S. D. & R. Y. Morita:
Effect of cultural conditions on growth of *Beggiatoa. J. Bact.*, 88:1755, 1964.
10. Classen, P. W:
"A Possible New Source of Food Supply," *Scientific Monthly*, 9:179-185, 1919.
11. Conant, J. B. & A. H. Blatt:
The Chemistry of Organic Compounds, Macmillan Co., New York, 1952.
12. Craighead, J. J., F. C. Craighead Jr. & R. J. Davis:
A Field Guide to Rocky Mountain Wildflowers, Houghton Mifflin Co., Boston, Mass., 1963.
13. DeLisle, H. F:

Common Plants of The Southern California Mountains, Naturegraph Co., 1961.

14. Galtsoff, P. S., F. E. Lutz & P. S. Welch:
Culture Methods For Invertebrate Animals, Dover Publications, Inc., New York, 1969.

15. Gibbons, E:
Stalking The Wild Asparagus, David McKay Co., New York, 1962.

16. Harrington, H. D:
Edible Native Plants of The Rocky Mountains, University of New Mexico Press, Albuquerque, New Mexico, 1967.

17. Harris, E. E:
Wood saccharification. *Advances in Carbohydrate Chemistry*, Vol. IV, Academic Press, New York, 1949.

18. Holmgren, A. H. and J. L. Reveal:
Checklist of the Vascular Plants of the Intermountain Region, U. S. Forest Service Research Paper INT-32, Ogden, Utah, 1966.

19. Kingsbury, J. M:
Poisonous Plants of The United States and Canada, Prentice-Hall, Inc., Englewood Cliffs, New Jersey, 1964.

20. Kirk, R. E. & D. F. Othmer (eds):
Descriptions of protein and wood hydrolysis. *Encyclopedia of Chemical Technology*, Interscience, Vol. 1, pp. 272-274, Vol. 11, pp. 212-220, 1956.
– Aspects of mass cultivation of microorganisms for food. *Encyclopedia of Chemical Technology*, Second edition, Interscience, Vol. 1, pp. 649-662, Vol. 16, pp. 635-638.

21. Lamanna, C. & M. F. Mallette:
Cellular makeup of bacteria and methods of lysis. *Basic Bacteriology*, Williams & Wilkins Co., Baltimore, Md., pp. 72-75, 84-89, 1965.

22. Leister, M. M:
Plant parts usually discarded. "Salads in Winter," *Flower Grower* 50:15, Jan. 1963.

23. Meacham. C:
"Nature's Untilled Gardens," *National Park Magazine* 35: 8-10, May 1961.

24. Medsger, O. P:
Edible Wild Plants, Macmillan Co., New York, 1966.

25. Olsen, L. D:

Outdoor Survival Skills, Brigham Young University Press, Provo, Utah, 1967.

26. Pelczar, M. J. Jr. & R. D. Reid:
Microbiology, McGraw-Hill, New York, 1965.

27. Perez-Llano, G. A:
"Lichens Used as Food by Man," *The Botanical Review*, 10(1):33-36, Jan. 1944.

28. Robinson, T:
Chemical fractionation of plants. *The Organic Constituents of Higher Plants*, Burgess Publishing Co., Minneapolis, Minn., 1967.

29. Rorty, J. & N. P. Norman:
Selenium absorbers. *Tomorrow's Food*, Prentice-Hall, Inc., New York, 1947.

30. Salle, A. J:
Composition of microorganisms. *Fundamental Principles of Bacteriology*, sixth edition, McGraw-Hill, New York, 1967.

31. Sarles, W. B. et al:
Microbiology, second edition, Harper & Sons, New York, 1956.

32. Smith, A. H:
The Mushroom Hunter's Field Guide, The University of Michigan Press, Ann Arbor, Michigan, 1967.

33. Sweet, M:
Common Edible and Useful Plants of The West, Naturegraph Co., Healdsburg, Calif., 1962.

34. Sweetman, B. J:
Amino acid composition of wool, *Text. Res. J.*, 37(10), pp. 834-844, 1967.

35. Thomas, W. S:
Field Book of Common Gilled Mushrooms, G. P. Putnam's Sons, New York, 1928.

36. Treadwell, F. P. & W. T. Hall:
Tests for selenium. *Analytical Chemistry*, Vol. I, J. Wiley & Sons, Inc., New York, pp. 509-510, 1959.

37. Umbreit, W. W:
The comparative physiology of autotrophic bacteria. *Bact. Rev.*, 26:145, 1962.

38. Welsh, S. L., M. Treshow & G. Moore:
Common Utah Plants, Brigham Young University Press, Provo, Utah, 1965.

39. Winogradsky, S. N:
"Concerning sulfur bacteria," *Botanische Zeitung*, XLV (1887), 489-507.
– "Investigations of Nitrifying Organisms," *Annales de l'Institut Pasteur*, IV (1890), 213-231.

INDEX

Abies spp, 9, 83
Abrin, 10, 96
Abronia latifolia, 9
Abrus precatorius, 9
Acer, 10
Acid(s), 34, 78, 99, 113, 123, 127, 128, 130, 134
Acorns, 92
Acorus calamus, 10
Adam's needle, 120
Adder's tongue, 48
Aesculin, 131
Aesculus californica, 10
Agaricus, 11-12
Agastache, 12, Pl. 1
Agave, 12
Agent, reducing, 129
Agoseris aurantiaca, 13
Agropyron, 13
Agrostis spp., 13
Ague tree, 101
Alcohol, 132
Alexanders, 105
Alfalfa, 92
Alfileria, 47
Algae, 122, 133-136
Alisma spp., 13
Alkaloid, heart depressant, 49
Alkaloids, 20, 66, 76, 79
Allium spp., 14
Allspice, 24
Alsine, 108
Alum root, 56
Amanita, 8, 14
Amaranthus, 8, 15
Amelanchier spp., 15, Pl. 7
American beech, 48
American bistort, 86
American brooklime, 117
American chestnut, 32
American coffee bean, 52
American cowslip, 28, 45
American licorice, 51
American lotus, 73
American nelumbo, 73
American sea rocket, 27
American skunk cabbage, 67
American speedwell, 117
American watercress, 31
American wayfaring tree, 118
Ammonia, 136
Amole, 36
Amphicarpa monoica, 16
Amsinckia tessellata, 16
Anagallis arvensis, 16
Analyses, food, 135
Anemonella thalictroides, 16
Anemopsis californica, 16
Animals, test, 133, 135
Anise-scented goldenrod, 106
Annona, 16-17
Annual, 8
Anthelmintic, 34
Anthemis nobilis, 17
Apios, 17
Apium graveolens, 17
Apocynum cannabinum, 17
Appalachian tea, 58, 118
Apples, 68
Apples, wild, 68
Apricots, 89
Aqua regia, 99
Aquilegia spp., 18
Arctium, 18
Arctostaphylos spp., 18
Argentina, 88
Arisaema triphyllum, 18
Arizona mesquite, 89
Armillaria, 18-19
Armoracia armoracia, 96
Arrowgrass, 113, 131
Arrowhead(s), 98, 118
Arrowleaf balsamroot, 23
Artemisia, 19-20
Arundinaria gigantea, 20
Asarum canadense, 20
Ascaridole, 34
Asclepias, 20-21, Pl. 1
Ash species, 49
Asimina triloba, 22

Asparagus, 10, 21, 22, 24, 30, 35, 46, 86, 91, 104
Asparagus officinalis, 22
Aspens, 87
Aspirator, 128
Asters, 22
Aster spp., 22
Astragalus, 22
Atenia gairdneri, 81
Atriplex, 22-23
Autotrophs, 136
Avena fatua, 23
Avens, 51
Azalea nudiflora, 94

B

Bacteria, 133-136
Bagasse, 129
Ballhead waterleaf, 57
Balsamorhiza, 23
Balsamroot, 23
Baptisia tinctoria, 23
Barbarea, 24, 121
Barberries, 24
Barnyard grass, 45
Barrel cactus, 48
Base, 123, 128
Basswoods, 112
Beans, 10, 95
Beans, green, 55
Bean vine, 81
Beargrass, 120
Bear's head, 56
Beaverbread, 90
Beebalm species, 71
Beefsteak morel, 54
Beefsteak mushroom, 48
[illegible]rasses, 13
[illegible]ne, 132
[illegible]*n aestivale*, 24
[illegible]*is*, 24, 68
Berries, 56, 60, 65, 68, 70, 72, 78, [illegible]83, 94, 100, 102-108, 114, [illegible] 119
Beta-cyano-L-alanine, 118
Betony, 108
Betula, 24-25
Beverage(s), 36, 38, 42, 43, 70, 79, 84, 90, 99, 116
Biennial, 9
Biennial wormwood, 19
Bigroot, 23
Bilberries, 116
Bilsted, 65
Birches, 25
Birdbills, 45
Bird's-foot cliff brake, 80
Bird's-foot fern, 80
Birthroot, 114
Biscuitroots, 42, 65
Bistort, 86
Bistorta, 86
Bitter buffaloberry, 103
Bittercresses, 31
Bitterroot, 64
Bitterroot lewisia, 64
Black alder, 58
Blackberries, 97
Black birch, 24
Black gum, 74
Black haw, 118
Black locust, 95, 131
Black lovage, 105
Black medic, 70
Black morel, 71
Black mulberry, 72
Black nightshade, 106
Black persimmon, 44
Black spruce, 83
Blazingstar, 64
Blister buttercup, 93
Blite goosefoot, 35
Blitum hastatum, 36
Blowball, 110
Blueberries, 116
Bluedicks, 26
Blue flax, 64
Blue-flowered lettuce, 61
Bluegrasses, 85
Blue mountain tea, 106
Blue pennycress, 111
Blue-sailors, 37
Bluewood, 40
Blushing venenarius, 14
Bog candle, 53

Bog myrtle, 72
Boisduvalia spp., 25
Bolander's yampa, 81
Boletellus russellii, 25
Boletes, 26
Boletinus cavipes, 25
Boletus, 25-26
Borage, 26
Borago officinalis, 26
Boxthorn, 66
Bracken fern, 91
Brake, 91
Branching phacelia, 81
Brassica spp., 26
Bread, 13, 27
Brick cap, 73
Brigham teas, 46
Bristly greenbrier, 104
Broad-leaved fireweed, 46
Broccoli, 24
Brodiaea, 26-27
Broomrapes, 76
Brussels sprouts, 53
Buckbrushes, 33
Buckthorn plantain, 84
Buffalo pea, 22
Bulrushes, 101
Bumelia spp., 27
Bunchberry, 41
Burreeds, 107
Bush chinquapin, 32
Bush cinquefoil, 88
Bush morning glory, 59
Bush oceanspray, 56
Bush rock spiraea, 56
Buttercups, 93
Butterfly tulip, 28
Butterfly weed, 21
Button snakeroot, 64

C

Cabbage palm, 98
Cabbage tree, 98
Caches, 16, 74
Cactaceae, 27
Cactus, 27
Caesar's mushroom, 14
Cakes, 59, 65, 95
Cakile edentula, 27
Calamus, 10
Calbovista subsculpta, 27
Calcium, 78
Calendula officinalis, 27
California bay tree, 115
California buckeye, 11
California fan palm, 119
California holly, 56
California juniper, 60
California laurel, 115
Calla palustris, 27
Calvatia, 29
Camash, 29
Camassia, 29
Campanula, 29
Canada lettuce, 61
Canadian buffaloberry, 103
Canary mushroom, 112
Candy, 10, 43, 48, 69
Cannabis sativa, 29
Canoe birch, 25
Cantharellus, 30
Capers, 28, 114
Capsella bursa-pastoris, 30
Carbohydrates, 92, 131-134
Carbonate, 136
Carbon, decolorizing, 129
Cardamine, 31
Carduus spp., 38
Carex spp., 31
Carica papaya, 31
Carnegiea gigantea, 31
Carolina buckthorn, 94
Carpophoroids, 94
Carrageen moss, 37
Carrion flower, 105
Carrots, 4, 7, 8, 43, 80
Carum gairdneri, 81
Carya spp., 31
Cassena, 58
Cassia occidentalis, 32
Castanea, 32
Castanopsis, 32
Catathelasma imperialis, 32
Catkins, 87
Catnip, 73

Cattails, 1, 5, 114, 125, 126
Ceanothus, 32-33, Pl. 7
Cedrus spp., 83
Cellulose, 129-131
Celtis, 33
Century plant, 12
Ceratonia siliqua, 33
Cercis spp., 33
Cereus giganteus, 33
Cetraria islandica, 34
Chaetochloa, 34
Chamaenerion angustifolium, 46
Chamomile, 17
Chanterelle, 30
Chaparral tea, 42
Checkerberry, 50
Cheese, cottage, 105
Cheeses, 68
Cheeseweed, 68
Chenopodium, 34-36
Cherries, 89
Chestnut, 32
Chickweed, 108
Chicory, 37, 121
Chicory lettuc, 61
Chicot, 51
Chile tarweed, 67
Chimaphila umbellata, 36
China brier, 105
Chinquapin, 32
Chiogenes hispidula, 36
Chlorate, 99
Chloride, 99, 129
Chloroform, 132
Chlorogalum pomeridianum, 36
Chlorophyllum molybdites, 37
Chocolate, 51
Chokecherries, 89-90
Chollas, 76
Chondrus crispus, 37
Christmas berry, 56
Chrysanthemum, 37, 110
Chrysobotrya aurea, 95
Chrysothamnus spp., 37
Chufas, 43
Cichorium intybus, 37, 121
Cicuta douglasii, 75, 104, 122
Cinnamon cortinarius, 41
Cinquefoil, 88
Cirsium, 38, Pl. 1
Cistus spp., 38
City goosefoot, 36
Cladonia rangiferina, 38
Clarkias, 38
Clarkia spp., 38
Clavariadelphus, 39
Clavaria spp., 38
Clay, 106
Claytonia, 39, 71
Cleome, 39
Clitocybe, 39
Clitopilus, 39-40
Clotbur, 18, 120
Clove bush, 95
Clovers, 113
Cluster lily, 26
Coats, wool, 127
Cochlearia armoracia, 96
Cocklebur, 120
Cockspur, 45
Coffee, 13, 18, 22, 32, 37, 43, 52, 53, 60, 92, 103, 111, 113, 114
Coffee senna, 32
Coffeeweed, 32
Collections, microbiological, 136-137
Collybia, 40
Columbines, 18
Commelina communis, 40
Common breadroot, 91
Common burdock, 18
Common chickweed, 108
Common goosefoot, 34
Common gourd, 42
Common licorice, 51
Common milkweed, 21
Common orange daylily, 55
Common plantain, 85
Common portulaca, 88
Common reed grass, 82
Common sagebrush, 19
Common winter cress, 24
Compass plant, 103
Compounds, nitrogen, 134
Comptonia peregrina, 72
Condalia obovata, 40
Condiment, 63
Coneflower, 93
Coontie, 121

Copper, 99
Coprinus, 40
Coral fungus, 56
Coral mushrooms, 38
Cornel, 41
Cornelian cherry, 41
Corn on the cob, 115
Corn poppy, 79
Cornsalads, 117
Corn stalks, 129
Cornus, 41
Cortinarius, 41
Corylus spp., 41
Coryphantha, 41
Cottonwoods, 87
Cow parsnip, 55
Cowslip, 18, 88
Costmary, 37
Crab apples, 68
Cranberry tree, 118
Crataegus spp., 41
Craterellus, 42
Creambush, 56
Creeping barberry, 68
Creeping bellflower, 29
Creeping Charlie, 74
Creeping snowberry, 36
Creeping wintergreen, 50
Crepis spp., 42
Crinkleroot, 43
Crops, aquatic, 126
Croton corymbulosus, 42
Crowfoot species, 93
Cuckoo flower, 31
Cucumber(s), 26, 69, 100
Cucurbita foetidissima, 42
Curled mallow, 69
Currants, 95
Cursed crowfoot, 93
Custard apple, 22, 31
Custards, 37
Cyanide, 68, 89, 90, 113
Cymopterus spp., 43
Cynoglossum grande, 43
Cyperus spp., 43
Cystine, 129

D

Dahoon holly, 58
Dalea terminalis, 43
Dandelions, 110, 121, 126
Darling plum, 93
Date plum, 44
Datil yucca, 120
Daucus, 43
Dayflower, 40
Daylily, 55
Deadly nightshade, 106
Deer grass, 94
Deertongue, 49
Delicious lactarius, 60
Dentaria spp., 43
Dentinum, 43
Deschampsia spp., 44
Descurainia, 44
Desert parsleys, 65
Desert thorn, 66
Devilsclaw, 69
Devil's claw, 102
D-fructose, 130
D-galactose, 130
D-glucose, 130
Digger pine, 84
Diospyros, 44
Dioxane, 132
Dioxide, carbon, 136
Dioxide, sulfur, 99
Disporum trachycarpum, 44
Distillation, 133
D-mannose, 130
Docks, 97, 121
Dodecatheon hendersonii, 44
Dogbane, 17
Dogtooth violet, 48
Double-claw, 69
Douglas fir, 90
Douglas spruce, 90
Dracocephalum nuttallii, 45
Dragonhead, 45
Drink(s), 10, 12, 26, 36, 38, 51, 52, 58, 76, 84, 93, 97, 110, 113
Duckweeds, 62
Dwarf cornel, 41
D-xylose, 130

E

Eagle fern, 91
Early morel, 117
Earth almonds, 43
Earthworms, 123
Echinocereus, 45
Echinochloa crusgalli, 45
Elaeagnus argentea, 45
Elderberries, 100
Elders, 100
Electrolyte, 123
Eleocharis palustris, 46
Elm mushroom, 85
Elymus spp., 46
Emetic, 48
Encinilla, 42
Enzyme, 128
Ephedra spp., 46
Epilobium, 46, Pl. 1
Equisetum, 46-47, Pl. 2
Ericaceae, 62
Eriocoma, 77
Eriodictyon californicum, 47
Eriogonum spp., 47, Pl. 3
Erodium cicutarium, 47
Erythronium, 47-48
Ether, 132
Evening primroses, 75
Evergreen winterberry, 58
Extractions, 132

F

Fagus, 48
Fairybells, 44
Fairy ring mushroom, 69
False dandelion, 13
False Solomon's seals, 104
Family, cactus, 27
Family, cattail, 114
Family, grass, 52
Family, mint, 60
Family, parsley, 65
Family, pine, 83
Fanweed, 111
Fasciola hepatica, 9
Fat(s), 19, 92, 131, 133, 134
Fava bean, 118
Fawn lily, 48
Feathers, 127
Fennel, 49
Fermentation, 119, 133
Ferocactus, 48
Fescue grasses, 48
Festuca spp., 48
Fiddleneck, 16
Field horsetail, 46
Field mushroom, 11
Field pennycress, 111
Filaree, 47
Filtration, 133
Fingernails, 127
Fireweed, 46
Fir species, 9
Fishweeds, 88
Fistulina hepatica, 48
Fivefingers, 88
Flammulina, 48
Flatsedges, 43
Flavorings, 12, 15, 19, 26, 30, 38, 49, 58, 59, 68, 69, 71, 72, 76, 81, 96, 98, 99, 101, 105, 110, 112
Fleecy lactarius, 61
Flies, 123
Flixweeds, 44
Florence fennel, 49
Florida arrowroot, 121
Flour, 13, 15, 25, 27, 34, 45, 47, 67, 74, 88, 89, 96, 101, 104, 115, 121, 125
Flukes, liver, 9
Fly agaric, 14
Foeniculum, 49
Forcing, 121
Foxtail grass, 34
Fragaria spp., 49, Pl. 2
Fragrant giant hyssop, 12
Frasera speciosa, 49
Fraxinus spp., 49
Fremont goosefoot, 35
Fritillaria, 49
Fritillarys, 49
Fritters, 95, 119
Fruit(s), 50, 51, 59, 69, 72, 74, 76, 77,

79, 80, 82, 85, 93-96, 111, 115 117, 120
Fungi, 52, 134
Fungicides, 132
Fungus, 107
Fuscoboletinus, 50

G

Galera tenera, 50
Galinsoga, 50
Galinsoga, 50
Garden burnet, 100
Garden nightshade, 106
Garget, 83
Garnish, 63
Gaultheria, 50
Gayfeather, 64
Gaylussacia spp., 51
Geum rivale, 51
Giant cactus, 33
Giant chinquapin, 32
Giant frasera, 49
Giant hyssop, 12
Giant puffball, 29
Gill-over-the-ground, 74
Ginger, 20, 31
Ginseng, 79
Glassworts, 98
Glechoma hederacea, 74
Gleditsia triacanthos, 51
Glistening coprinus, 40
Globemallow, 59
Glyceria spp., 51
Glycoalkaloids, 105
Glycoside(s), 20, 33, 68, 93, 118, 131, 132
Glycyrrhiza, 51, Pl. 7
Goat nut, 103
Goatsbeard, 112
Godetia biloba, 51
Golden club, 77
Golden gooseberry, 95
Golden purslane, 88
Goldenrods, 106-107
Gomphidius, 51
Good-King-Henry, 35
Gooseberries, 95
Goosefoot species, 36
Goosegrass, 88, 113
Graminae, 52
Grand hound's tongue, 43
Grapes, 119
Grayling, 30
Greasewood, 100
Great basin orogenia, 77
Great burdock, 18
Great duckweeds, 108
Green amaranth, 15
Green arrow arum, 80
Greenbrier, 104
Green gentian, 49
Green hedgehog cactus, 48
Green russula, 97
Greens, edible, 122
Grindelia squarrosa, 52, Pl. 7
Gromwell, 65
Grossularia spp., 95
Ground-cherry species, 82
Ground hemlocks, 111
Ground ivy, 74
Groundnut, 17
Ground plum, 22
Gum, 13, 17, 21, 59, 65, 67, 83, 84, 103, 112
Gumweed, 52
Gymnocladus dioica, 52
Gypsum, 136
Gyrodon merulioides, 52
Gyroporus, 52-53

H

Habenaria dilatata, 53
Hackberry, 33
Hair, 127
Hairgrasses, 44
Hairleaf water crowfoot, 92
Hairy vetch, 118
Half-free morel, 72
Harebell, 29
Harsh sidalcea, 103
Hawk's beards, 42
Hawthorns, 41

Hazelnut species, 41
Hedge mustards, 103
Hedysarum, 53
Helianthus, 53-54, Pl. 5
Helvella, 54
Hemerocallis, 55
Hemlocks, 114
Hemp, 30
Henderson shootingstar, 45
Hen of the woods, 87
Heracleum, 55, Pl. 5
Herbicides, 132
Hercules-parsnip, 55
Hericium, 55
Heteromeles arbutifolia, 56
Heuchera micrantha, 56
Hickory trees, 31
High-bush cranberry, 118
High mallow, 69
Hippuris, 56
Hips, rose, 96
Hobblebush, 118
Hog peanut, 16
Holly grape, 68
Hollyhocks, 59
Holodiscus, 56
Honey, 11
Honeydew, 87
Honey locust, 51
Honey mesquite, 89
Honey mushroom, 19
Honeypod, 89
Honey shucks, 51
Honeysuckle, 65
Hooker's balsamroot, 23
Hooves, 127
Horehound, 69
Horn of plenty, 42
Horns, 127
Horsechestnuts, 131
Horsemint(s), 12, 71
Horseradish, 96
Horsetails, 47
Hottentot fig, 70
Huckleberries, 51
Humins, 129
Husk-tomatoes, 82
Hydnum, 56-57
Hydrolysis, 127, 129
Hydrophyllum, 57
Hydroquinone, 120
Hydroxide(s), 92, 128
Hydroxylamine hydrochloride, 99
Hygrophorus, 57
Hypericin, 57
Hypericum, 57-58
Hypholoma, 58
Hyssop, 58
Hyssopus officinalis, 58

I

Iceland moss, 34
Ice plant, 70
Ilex, 58
Iliamna rivularis, 59, Pl. 5
Impatiens biflora, 59
Indian black drink, 58
Indian breadroot, 91
Indian cherry, 94
Indian cucumber root, 69
Indian hemp, 17
Indian lettuce, 71
Indian millet, 77
Indian mountain-rice, 77
Indianpaint, 65
Indian pond lily, 74
Indian potato, 77
Indian rhubarb, 80
Indian rice, 121
Indian ricegrass, 77
Indian turnip, 18, 91
Indian wheat, 85
Indigo bush, 43
Infusions, 135
Inky cap, 40
Insecticides, 9, 122, 132
Insects, 122-123
Ipomoea, 59
Irish moss, 37
Isabella wood, 81
Isomaltose, 130
Isothiocyanate, allyl, 111
Ivory plum, 36

J

Jack-in-the-pulpit, 18, 27
Jacob's ladder, 104
Jams, 15, 16, 18, 86
Japanese fleeceflower, 86
Japanese knotweed, 86
Japanese persimmon, 44
Jellies, 15, 16, 18, 24, 68, 81
Jerusalem artichoke, 54
Jerusalemoak goosefoot, 35
Jewel weed, 59
Jointfirs, 46
Jojoba, 103
Judas tree, 33
Juglans spp., 59
Juice, 21, 58, 110
Junipers, 60
Juniperus, 59-60

K

Kentucky coffee tree, 52
Keratins, 127
King's crowns, 102
Klamath weeds, 57
Knotweed(s), 86-87

L

Labiatae, 60, Pl. 1
Labrador tea, 62
Laccaria, 60
Lactarius, 60-61
Lactuca, 61, Pl. 2
Lambsquarters, 34-35
Laportea canadensis, 62
Large cane, 20
Large flowered trillium, 114
Larger ground plum, 22
Larger withe rod, 118
Large tupelo, 74
Larix spp., 83
Larvae, mosquito, 123
Lateiporus sulphureus, 62
Laurel, 71
Laurel magnolia, 68
Layers, slime, 126
Lead, 132
Leather, 127
Ledum groenlandicum, 62
Lemaireocereus thurberi, 62
Lemna spp., 62
Lemonade, 24, 94
Lemon balm, 70
Lentinus lepideus, 62
Leopard lily, 49
Lepargyraea argentea, 102
Lepidium, 63
Lepiota, 63
Lepista, 63
Lettuce, 8
Lettuce saxifrage, 101
Leucoagaricus, 63
Levisticum officinale, 63
Lewisia rediviva, 64
Liatris punctata, 64
Lichens, 108
Licorice, 77, 116
Licorice root, 53
Lilies, 64
Lilium spp., 64
Lily species, 64
Limacella lenticularis, 64
Lime-flowers, 112
Limestone, 136
Limnia perfoliata, 71
Limnobotrya spp., 95
Linum lewisii, 64, Pl. 5
Liquidambar styraciflua, 65
Lithospermum, 65
Littleflower quickweed, 50
Liverberry, 109
Lobed godetia, 51
Locusts, 123
Locust tree, 95
Lomatium spp., 65
Long-stalked greenbrier, 105
Lonicera, 65
Lovage, 63
Lupinus spp., 65
Lycium, 66
Lycoperdon spp., 67

Lye, 104
Lygodesmia, 67
Lysichitum americanum, 67

M

Machine, grinding, 132
Madia sativa, 67
Magnolia virginiana, 68
Mahonia repens, 68
Mallow, 68
Malus spp., 68
Malva, 68-69
Mandarin, 44
Mandrake, 85
Mannagrasses, 51
Man-of-the-earth, 59
Manzanitas, 18
Mapleleaf goosefoot, 35
Maplemallow, 59
Maples, 10
Marasmius oreades, 69
Marestail, 56
Marigold, 27
Marijuana, 29-30
Mariposa lily, 28
Marjorams, 76
Marrubium vulgare, 69
Marsh cress, 96
Marshmallows, 82
Marsh marigold, 28
Martynia louisiana, 69
Masterwort, 55
Maté, 58
Matrimony vine, 66
Matter, nitrogenous, 91
Matteuccia struthiopteris, 75
Maximilian sunflower, 54
Mayapple, 85
Maypop, 79
Meadow beauty, 94
Meadowbright, 28
Meadow cress, 31
Meadow hygrophorus, 57
Meadow salsify, 112
Meal, 11, 13, 15-19, 23, 30, 38, 39, 42, 44, 45, 56, 57, 64, 65, 74-76, 79, 80, 84, 88, 90-93, 97, 99, 102, 103, 107-109, 113, 117, 119, 120
Meat, 127
Mecha-meck, 59
Medeola virginiana, 69
Medicago lupulina, 70
Melissa officinalis, 70
Mentha spp., 70, Pl. 1
Mentzelia spp., 70
Mesembryanthemum, 70
Mesquite, 89
Mexican persimmon, 44
Micranthes micranthidifolia, 101
Microbes, 134-135
Microorganisms, 7, 131, 134-136
Milk, 27, 105
Milkvetch, 22
Milkweeds, 20
Mimulus, 70
Minerals, 131
Miner's lettuce, 71
Mint geranium, 37
Mint species, 70
Missouri goldenrod, 106
Mitchella repens, 70
Molasses, 33
Molds, 133-135
Molybdenum, 132
Monarda spp., 71
Monkeyflowers, 70
Monolepis, 71
Montia, 71
Monument plant, 49
Mooseberry, 118
Morchella, 71-72
Morel, 72
Mormon teas, 46
Morus, 72
Moths, 123
Mountain ash, 107
Mountain dandelion, 13
Mountain hedge, 79
Mountain holly, 68
Mountain hollyhock, 59
Mountain lettuce, 101
Mountain lover, 78
Mountain ricegrass, 77
Mountainspray, 56

Mountain tea, 50
Mud, 122
Mugwort, 20
Mulesears, 119
Munz black-headed cactus, 45
Mush, 45, 47
Mustard, 26
Mustard species, 26
Mycelia, 134
Mycena galericulata, 72
Myrica, 72
Myrrhis odorata, 72
Myrtle boxleaf, 78

N

Naematoloma, 73
Nannyberry, 118
Narrow-leaved plantain, 84
Nasturtium, 114
Nasturtium officinale, 96
Naucoria semiorbicularis, 73
Navajo teas, 111
Navarretia spp., 73
Nectar, 95
Nelumbo lutea, 73
Nemexia, 104
Neomammilaria applanata, 73
Nepeta, 73
Netleaf hackberry, 33
Nettles, 116
New Jersey teas, 32
Nightshade, 106
Nipple cactus, 73
Nitrate(s), 8, 15, 26, 34, 38, 39, 45, 67, 71, 99, 100, 107, 108, 132, 135
Nitrogen, 136
Nonesuch, 70
Northern sweetvetch, 53
Nuphar, 74
Nutrients, 133
Nutgrasses, 43
Nuts, 16, 31, 41, 48, 59, 84, 103
Nuttall monolepis, 71
Nymphaea polysepalum, 74
Nyssa, 74

O

Oak-leaf geranium, 80
Oak-loving mushroom, 40
Oaks, 92
Oceanspray, 56
Oenanthe sarmentosa, 75
Oenothera, 75
Oil(s), 19, 24, 26, 53, 74, 79, 92, 93, 111
Old man of the woods, 109
Omphalia campanella, 75
Onions, 14
Onoclea, 75
Opium poppy, 79
Opuntia spp., 76
Orange amanita, 14
Orange-brown lactarius, 61
Orange clitocybe, 39
Orange honeysuckle, 65
Orchidaceae, 76
Orchids, 53
Oregon boxwood, 78
Oregon grapes, 24, 68
Oregon myrtle, 115
Organ pipe cactus, 62
Origanum spp., 76
Ornithogalum umbellatum, 76
Orobanche spp., 76
Orogenia linearifolia, 77
Orontium aquaticum, 77
Orpines, 102
Oryzopsis, 77
Osmorhiza spp., 77, Pl. 8
Ostrich fern, 76
Oxalate(s), 18, 28, 67, 78, 83, 88, 97, 98, 100, 110
Oxalis, 78
Oyster mushroom, 85
Oysterplant, 112

P

Pachystima myrsinites, 78

Pacific trillium, 114
Paint, 39
Pale wolfberry, 66
Palmetto, 98
Palmer's amaranth, 15
Panaeolus retirugis, 79
Panicum crusgalli, 45
Panus strigosus, 79
Papaver, 79
Papaw, 22, 31
Papaya, 31
Paperbirch, 25
Paraguay tea, 58
Parasol mushroom, 63
Parsnip(s), 28, 77
Partridgeberry, 60, 70
Passiflora, 79
Passion flower, 79
Passion vine, 79
Patata, 71
Patties, 102
Paxillus, 79
Peaches, 89
Peas, 7
Pedicularis lanata, 80
Pediomelum esculentum, 90
Pelargonium capitatum, 80
Pellaea ornithopus, 80
Peltandra virginica, 80
Peltiphyllum peltatum, 80
Peniocereus greggi, 80
Pennsylvania bitter cress, 31
Pentoses, 130
Pepper, 96
Peppergrass, 63
Pepperidge, 74
Pepperleaf, 96
Pepperroots, 43
Pepperweeds, 63
Peptides, 128
Poptones, 134
Peraphyllum ramosissimum, 81
Perfoliate bellwort, 116
Perideridia, 81
Peritoma serrulatum, 39
Persia borbonia, 81
Persimmon, 44
Petroleum, 135
Phacelia ramosissima, 81
Phaseolus, 81
Pholiota, 82
Phosphate, potassium, 135
Photinia arbutifolia, 56
Phragmites communis, 82
Phylloporus rhodoxanthus, 82
Physalis spp., 82
Phytolacca, 63, 82, 121
Phytotoxin, 95-96
Picea, 83
Pickles, 22, 32, 47, 69, 83, 88, 93, 94, 98, 104
Pickleweeds, 98
Pies, 50, 83, 86, 100, 106
Pigeonberry, 83
Pigment, toxic, 58
Pig's ears, 30
Pigweed, 34
Pilotweed, 103
Pimbina, 118
Pimple mallow, 28
Pinaceae, 83
Pine cone mushroom, 109
Pinenuts, 84
Pink cleome, 39
Pinkets, 47
Pink milkweed, 21
Pinole, 67
Pinus, 83-84
Pinxter flower, 94
Pipsissewa, 36
Pitseed goosefoot, 35
Plantago, 84-85
Plant, poisonous, 131
Plants, aquatic, 9, 122
Plants, edible, 3, 5, 7
Pleurotis, 85
Plums, wild, 89
Pluteus, 85
Pluteus, 85
Poa spp., 85
Podgrass, 113
Podophyllin, 85
Podophyllum peltatum, 85
Poisonberry, 106
Poisoning, 8
Poison ivy, 132, Pl. 8
Poison milkweed, 21
Poison oak, 132

Poke, 83
Pokeweed, 83, 121
Pollen, 102, 115
Pollution, 122
Polygonatum biflorum, 86
Polygonum, 86-87
Polyozellus multiplex, 87
Polypeptides, 134
Polypilus, 87
Polyporus sulphureus, 87
Polysaccharides, 130, 136
Pond apple, 16
Pondweeds, 88
Popcorn, 74
Poplars, 87
Populus spp., 87, Pl. 6
Porridge, 18
Portulaca, 88
Potamogeton spp., 88
Potato(es), 17, 64, 76, 81, 105, 106, 112, 125
Potatoes, sweet, 73
Potentilla, 88, Pl. 3
Poterium, 100
Potherb(s), 20, 24, 26, 34, 35, 39, 50, 61, 62, 66, 71, 75, 84, 96, 98, 102, 103, 106, 107, 110, 113, 114, 116, 117
Pot-pourri, 19
Poverty weed, 71
Prarie apple, 91
Prarie bean, 81
Prarie coneflower, 93
Prarie flax, 64
Prarie potato, 91
Prarie sunflower, 54
Precatory bean, 10
Preserves, 33, 50, 74, 79, 100, 102, 106, 116, 118
Prickly lettuce, 61
Prickly pears, 76
Primula veris, 88
Prince's pine, 36
Proboscoidea lousianica, 69
Prosopis, 89, 109
Prostrate amaranth, 15
Protein(s), 10, 26, 92, 102, 113, 122, 123, 127, 131, 133, 134
Protoanemonin, 28, 93
Protozoa, 122, 135, 136
Prune mushroom, 40
Prunus, 89-90, Pl. 6
Psathyrella disseminata, 90
Pseudotsuga, 83, 90
Psilocybe foenisecii, 90
Psoralea, 90-91
Puccoon, 65
Puddings, 37, 86, 106
Puffballs, 29, 67
Purple haw, 40
Purple oyster plant, 112
Purple-spot fritillaria, 49
Purslane, 88
Pusley, 88
Pussyfoot, 57
Pyrus spp., 92

Q

Quackgrass, 13
Queen Anne's lace, 43
Queen's crowns, 102
Quercus spp., 92
Quickweed, 50

R

Rabbitberry, 103
Rabbitbrushes, 37
Rabbit thorn, 66
Radicula, 96
Radishes, 8
Rainbow cactus, 45
Raisins, 15
Rampion, 29
Ranunculin, 93
Ranunculus, 92-93
Raspberries, 97
Ratibida columnifera, 93
Red amaranth, 15
Red bay, 81
Redberry, 94
Redbud, 33
Red cedar, 60

Red elm, 115
Red-flowered torch cactus, 45
Red gum, 65
Red hypholoma, 58
Red ironwood, 93
Red maple, 10
Red mulberry, 72
Red-osier dogwood, 41
Red poppy, 79
Redroot, 15
Red spruce, 83
Redtop grasses, 13
Red twinberry, 65
Red willow, 41
Red willow-weed, 46
Reed, 82
Reindeer moss, 38
Resinoids, 20, 62
Resinweed, 52, 103
Reynosia latifolia, 93
Rhamnus, 94
Rhexia virginica, 94
Rhododendron nudiflorum, 94
Rhodophyllus abortivus, 94
Rhodotorula gracilis, 134
Rhubarb, 38, 78, 86
Rhus, 94-95
Ribes, 95, Pl. 3
Rice, 49
Ricin, 96
Roads, 126
Robinia pseudo-acacia, 95
Rockrose, 64
Rock-spirea, 56
Rock tripes, 116
Rocky Mountain beeplant, 39
Rocky Mountain cowlily, 74
Rocky Mountain juniper, 60
Rocky Mountain maple, 10
Root mushroom, 40
Roots, 121
Rorippa, 96, Pl. 3
Rosary pea, 10
Rosa spp., 96, Pl. 4
Roses, 96
Rough-leaved hackberry, 33
Rough-leaved mallow, 68
Round-leaved psoralea, 91
Round-leaved watercress, 31
Rowan tree, 107
Rozites caperata, 97
Rubus spp., 97, Pl. 4
Rue, 98
Rue anemone, 16
Rumex, 97, 121, Pl. 4
Rushpink, 67
Russet buffaloberry, 103
Russian tarragon, 19
Russian thistle, 98
Russula, 97
Rusts, 107
Rutabega, 55
Ruta graveolens, 98

S

Sabal palmetto, 98
Sabina, 59-60
Saccharomyces cerevisiae, 134
Sagebrush, 19
Sages, 99
Sagittaria spp., 98
Sahuaro, 31
Salad burnet, 100
Salad(s), 10, 16, 19, 24, 26-29, 31, 33, 34, 37, 55, 61, 62, 68, 70, 72, 75, 78, 84, 86, 88, 94, 96-98, 100-103, 105, 107, 108, 110, 113, 117, 119
Salicornia spp., 98
Salicylate, methyl, 24
Salix spp., 98
Sallal, 50
Salsify, 112
Salsola, 98
Salt, 23, 55, 123, 128
Saltbushes, 23
Saltwort, 98
Salvia, 99-100
Sambucus spp., 100, Pl. 8
Samphires, 98
Sanguisorba minor, 100
Saponins, 33, 83, 103, 121
Sarcobatus vermiculatus, 100
Sarcoscypha coccinea, 101
Sassafras, 101

Sassafras, 101
Satureja, 101
Sauce(s), 66, 86, 118
Saxifraga peltata, 80, 101
Scaly lentinus, 63
Scaly pholiota, 82
Scarlet cup, 101
Scarlet pimpernel, 16
Scent bottle, 53
Schmaltzia, 94-95
Scirpus, 101-102
Sclerocactus whipplei, 102
Scleroderma aurantium, 102
Scoke, 83
Scouring rush, 46
Screw bean, 89, 109
Scurfpea, 90, 91
Scurvy grass, 24
Seacoast abronia, 9
Sea fig, 70
Sedges, 31
Sedum spp., 102, Pl. 4
Seeds, sesame, 79
Seeds, sunflower, 53
Sego lily, 28
Selenium, 22, 23, 52, 98, 132
Sensitive fern, 75
Serviceberries, 15
Shadscale saltbush, 23
Shaggy mane, 40
Sheathed amanitopsis, 15
Sheepsorrel, 97
Shepherdia, 102-103
Shepherd's purse, 30
Shoes, 127
Shore podgrass, 113
Showy milkweed, 21
Shrubby cinquefoil, 88
Sidalcea malvaeflora, 103
Silk, 127
Silkweed, 21
Silphium laciniatum, 103
Silt, 122
Silverberry, 45
Silver buffaloberry, 102
Silverleaf, 103
Silverleaf nightshade, 105
Silverweed, 88
Silverweed cinquefoil, 88
Simmondsia chinensis, 103
Sisymbrium spp., 103
Sitting cactus, 45
Sium, 104
Skeleton plant, 67
Skeleton weed, 67
Skin, 127
Skunkbush, 95
Skunkbush sumac, 95
Skunk cabbage, 110
Skunktop scurfpea, 90
Skunkweeds, 73
Slim amaranth, 15
Slimleaf goosefoot, 35
Slim stinging nettle, 116
Slippery elm, 115
Slippery jack, 109
Smallage, 17
Small-flowered nightshade, 106
Small Indian breadroot, 91
Small soapweed, 120
Smartweeds, 87
Smilacina spp., 104
Smilax, 104-105
Smooth horsetail, 47
Smooth sumac, 94
Smyrnium olusatrum, 105
Snakeweed, 86
Snowdrops, 77
Snow lily, 48
Snow mushroom, 54
Soapberry, 103
Soap plant, 36
Soapweed, 120
Softstem bulrush, 102
Solanum, 105-106
Solanums, 105
Solidago, 106-107
Solomon plumes, 104
Solvent, 132-133
Sonchus spp., 107
Sooty lactarius, 61
Sophia spp., 107
Sorbus, 107
Sorrels, 97
Soup(s), 32, 66, 68, 69, 88, 91, 101, 122
Sour cactus, 41
Sourgrass, 113

Sour gum, 74
Soursop, 16
Southern California Goldenrod, 106
Southern hackberry, 33
Southernwood, 19
Sowthistles, 107
Soybean, 133
Spanish bayonet, 120
Spanish dagger, 120
Sparassis radicata, 107
Sparganium spp., 107
Spathyema foetida, 110
Spatterdock, 74
Speedwells, 117
Sphaeralcea rivularis, 59
Spicebush, 24
Spiderflower, 39
Spikeprimroses, 15
Spikerush, 46
Spinach, 21, 28, 34-36, 78, 83, 111
Spiny stars, 41
Spiraea spp., 108
Spireas, 108
Spirodela spp., 108
Spotted ladysthumb, 87
Spotted touch-me-not, 59
Spring beauties, 39
Squashberry, 118
Squaw apple, 81
Squawbush, 95
Squaw currant, 95
Squawroot, 81
Stachys palustris, 108
Stagbush, 118
Starch, 77, 86, 91, 102, 114, 115, 121
Star-of-Bethlehem, 76
Star tulip, 28
Steinpilz, 26
Stellaria, 108
Stews, 69
Sticta glomulifera, 108
Stinging nettle, 116, 121
Stinkweed, 39
St. John's bread, 33
St. Johnsworts, 57
Stonecrops, 102
Storksbill, 47
Strawberries, 49
Strawberry blite, 35
Strawberry cactus, 45
Streptopus amplexifolius, 108
Strobilomyces, 109
Strombocarpa pubescens, 109
Stropharia ambigua, 109
Succory, 37
Sugarberry, 33
Sugar pine, 83
Sugar(s), 21, 77, 82, 83, 89, 91, 102, 105, 109, 129-131, 134, 135
Suillus, 109
Sulfate(s), 99, 132, 135
Sulfur, 136
Sulfur-milk lactarius, 61
Sulfur mushroom, 87
Sulfur shelf, 62
Sumacs, 94
Summer savory, 101
Sunflower(s), 53-54
Sun roses, 38
Swamp milkweed, 20
Sweet bay, 68, 81
Sweet birch, 24
Sweetbroom, 53
Sweet chestnut, 32
Sweet cicely(s), 72, 77
Sweet fern, 72
Sweet flag, 10
Sweet gale, 72
Sweet goldenrod, 106
Sweet gum, 65
Sweetish lactarius, 61
Sweet potato cactus, 86
Sweetroots, 77
Sweetsop, 17
Sweet viburnum, 118
Swertia radiata, 49
Symplocarpus foetidus, 110
Syrup, 10, 21, 24, 31, 84, 101, 109

T

Tanacetum vulgare, 110
Tanglefoot, 118
Tannin, 92
Tansy, 110

Tansy mustards, 44
Tapioca, 116
Taraxacum, 110, 121
Tarragon, 19
Tarweed(s), 52, 67
Taw-kee, 77
Taxus spp., 111
Tea, 10, 17-20, 24, 25, 32, 34, 43, 46, 49, 50, 58, 60, 62, 69, 70-73, 80, 84, 88, 90, 93, 97, 101, 106, 108, 110-114, 116, 118
Teaberry, 50
Tea fern, 80
Tepary bean, 81
Termites, 122
Terpene, 34
Tetrahydrocannabinol, 30
Thelesperma, 111
Thiaminase, 47, 91
Thiamine, 47, 91
Thick-footed morel, 71
Thimbleberries, 97
Thiourea, 99
Thistle species, 38
Thlaspi, 111
Thymes, 112
Thymus spp., 112
Tiger lily, 49
Tilia spp., 112
Tobacco-root, 117
Toenails, 127
Togaria aurea, 112
Tomatilla, 66
Tomato, 105
Toothworts, 43
Tops, beet, 78
Tornillo, 109
Torrey wolfberry, 66
Torulopsis utilis, 134
Toxins, 8, 132
Toyon berry, 56
Tragopogon, 112, Pl. 6
Trees, pine, 5, 84
Trichloroethylene, 133
Tricholoma, 112-113
Trifolium spp., 113
Triglochin maritima, 113
Trilliums, 114
Trophies, 127
Tropillo, 105
True Solomon's seal, 86
True wood sorrel, 78
Trumpet of death, 42
Tryptophan, 127, 129
Tsuga spp., 83
Tuber starwort, 108
Tumblemustards, 103
Tumbleweed, 98
Tupelo, 74
Turkey peas, 77
Twinberry, 65, 70
Twisted-stalk, 108
Typha(ceae), 114, 125
Tyrosine, 129

U

Ulmus fulva, 115
Umbellularia californica, 115
Umbilicaria, 116
Umbrella plants, 47
Unicorn plant, 69
Urea, 129
Urine, 136
Urtica, 116, 121
Urticastrum divaricatum, 62
Utah aloe, 12
Uvularia perfoliata, 116

V

Vaccinium spp., 116
Vacuum, 128, 130, 133
Vagnera spp., 104
Valeriana edulis, 116
Valerians, 117
Velvet collybia, 40
Vermilion mushroom, 57
Veronica americana, 117
Verpa bohemica, 117
Vesicant, 132
Vetches, 118
Viburnum, 117-118
Vicia(s), 118-119

Viola spp., 119
Violets, 119
Violet wood sorrel, 78
Virginia tuckahoe, 80
Virginia waterleaf, 57
Vitamin(s), 44, 71, 84, 87, 96, 111, 116, 117, 131, 133, 134, 136
Vitis spp., 119
Volvaria bombycina, 119

W

Walnut species, 59
Wake robin, 114
Water arum, 27
Water avens, 51
Water beans, 73
Water buttercup, 92
Water chinquapin, 73
Watercress, 1, 9, 43, 96
Watergrass, 45
Water hemlock, 104, 122
Waterleaf (species), 57
Water montia, 71
Water oats, 121
Water-parsley, 75, 104
Waters, surface, 135
Washingtonia filifera, 119
Wastes, animal, 136
Wax currant, 95
Waxflower, 36
Waxy mushroom, 60
Weeds, 126
Western bistort, 86
Western hackberry, 33
Western junipers, 59, 60
Western scouringrush, 47
Western sweetvetch, 53
Western wintergreen, 50
Wheat, 20, 125
Wheatgrass, 13
Whiskers, 127
White bog orchid, 53
White horsenettle, 105
White mandarin, 109
White mulberry, 72
White pine, 84
Whorled mallow, 69
Whortleberries, 116
Wild bean, 17, 81
Wild calla, 27
Wild caraway, 81
Wild carrot, 43
Wild celery, 17
Wild coffees, 114
Wild flax, 64
Wild ginger, 20
Wild hyacinth, 26, 27, 29
Wild indigo, 24
Wild lemon, 85
Wild lettuce, 61
Wild licorice, 51
Wild oat, 23, 116
Wild olive, 74
Wild peanut, 16
Wild potato, 36, 105, 106
Wild potato vine, 59
Wild raisin, 118
Wild rice, 121
Wildrye species, 46
Wild spinach, 35
Willow herb, 46
Willows, 98
Willow-weeds, 46
Winterberry, 58
Winter cress, 24, 121
Wintergreen, 50
Winter mushroom, 48
Winter savory, 101
Wisterias, 119
Wisteria spp., 119
Withe rod, 118
Wolfberries, 66
Wood, 129, 130
Wood nettle, 62
Woolen-breeches, 57
Wormseed, 34
Wormseed goosefoot, 34
Wyethia spp., 119

X

Xanthium strumarium, 120

Y

Yampa, 81
Yaupon, 58
Yeast, 129, 130, 133-135
Yellow adder's tongue, 47
Yellow beeplant, 39
Yellowbell, 49
Yellow birch, 25
Yellow fritillary, 49
Yellow passion flower, 79
Yellow rocket, 24
Yellow rose, 88
Yellow skunkcabbage, 67
Yellow water cress, 96
Yellow waterlily, 74
Yerba, 58
Yerba mansa, 16
Yerba santa, 47
Yews, 111
Yucca, 120
Yuccas, 120

Z

Zamia, 121
Zinc, 99
Zizania, 121
Zygadenus, 29